Lecture Not
Mathematics

Edited by A. Dold and B. Eckmann

745

David K. Haley

Equational Compactness in Rings

With Applications to the Theory of Topological Rings

Springer-Verlag
Berlin Heidelberg New York 1979

Author

David K. Haley
Fakultät für Mathematik
und Informatik
Universität Mannheim
D-6800 Mannheim A 5

AMS Subject Classifications (1970): 02H15, 13J99, 13L05, 16A46, 16A80, 22A30

ISBN 3-540-09548-9 Springer-Verlag Berlin Heidelberg New York
ISBN 0-387-09548-9 Springer-Verlag New York Heidelberg Berlin

Library of Congress Cataloging in Publication Data
Haley, David K 1942-
Equational compactness in rings, with applications to the theory of topological rings.
(Lecture notes in mathematics ; v. 745)
Bibliography: p.
Includes index.
1. Commutative rings. 2. Associative rings. 3. Topological rings. I. Title. II. Series:
Lecture notes in mathematics (Berlin) ; v. 745.
QA3.L28 no. 745 [QA251.3] 510'.8s [512'.4] 79-20682
ISBN 0-387-09548-9

Printing and binding: Beltz Offsetdruck, Hemsbach/Bergstr.
2141/3140-543210

TABLE OF CONTENTS

INTRODUCTION

The intent of this treatise is to provide a conspectus of equational compactness in rings and those consequences of the theory of equational compactness yielding new insight into the older, classical subject of compact topological rings. A ring R (or more generally, any universal algebra) is said to be equationally compact if every system of algebraic equations is simultaneously solvable in R as soon as each finite subsystem is (all concepts will be made precise in Chapter I). It follows easily from the Tychonoff Product Theorem that a compact topological ring (or universal algebra) is equationally compact, and the concept furnishes in this way a kind of algebraic "approximation" of this topological algebraic property.

Equational compactness can appear in various disguises; so it is that the equationally compact Boolean algebras are just the complete ones [53], and the equationally compact abelian groups form the important class of algebraically compact abelian groups, introduced by I. Kaplansky some 25 years ago (see [1], [29], [34]). Indeed, it was this connection with abelian groups which in part motivated J. Mycielski to initiate the study of equational compactness in universal algebras [38]. Now abelian groups can be viewed as the equational class of zero rings, and of course Boolean algebras are just Boolean rings painted green, hence the study of equationally compact associative rings provides a common roof for the aforementioned classes of structures.

Although the applications of the theory of equationally compact rings so far seem to lie mainly in the area of compact topological rings and the pursuance of those applications is one of our central goals, we shall nevertheless refrain from any attempt to incorporate into this work a survey of the latter subject. The existing literature on compact topological rings is far too wide to do it justice and still remain within the framework of equational compactness, which forms the core of this exposition. In this connection we are looking forward to the appearance of a monograph which we have heard is in preparation by V.I. Arnautov et al in Novosibirsk promising to give a comprehensive survey of the literature on compact topological rings.

How, one may ask, can the introduction of a seemingly superficial concept, as equational compactness at first glance may appear to be, lead to any real contribution to the theory of topological rings, a theory which since the 1930's has been pursued and furthered by a number of renown mathematical minds? The answer lies in a single word: the ultraproduct - an algebraic construction which came into being not until the middle 1950's. It is precisely the ultraproduct which opens the door to the model theoretic setting of equational compactness as reflected, perhaps most notably, in the theorem of Weglorz [53, Theorem 4.1]. This theorem represents the most crucial model theoretic tool in our investigations. The connections will be made explicit in Chapter I where, in particular, the "structure topologies" are introduced and which are of central importance for our methods. We stress that these are not topological algebraic in essence but rather model theoretic, and represent a translation into topological terms of the model theoretic results of the theory of equationally compact universal algebras. A brief introduction in this direction will provide therefore an understanding for the model theoretic significance of these topologies as well as for their application potential.

I. Kaplansky [28] proved that a compact topological (Hausdorff) simple ring is finite, a result of considerable depth. The more ambitious question, whether there exists a compact topological ring containing an infinite simple subring, has, however, not been settled sofar, but is a question of obvious relevance when searching for necessary conditions for a ring to carry a compact topology compatible with its structure. This question and others in a similar vein are taken up in Chapter II and our attention is not restricted to "simplicity" alone. It turns out that minimum conditions - of one kind or another - provide a certain lever in attacking "compactification" questions of this nature. The main result is a very transparent characterization of quasi-compactifiable (i.e., being embeddable into a ring satisfying a weakened form of equational compactness) artinian rings, Theorem 4.13. In this class of rings this compactification property is in fact equivalent to equational compactness, although <u>not</u> to compactness itself (compact topological artinian rings are finite). The equivalence of quasi-compactifiability and equational compactness (even compactness) is demonstrated in simple rings satisfying certain additional minimum conditions (e.g., non-zero socle), and thus the general question posed at the beginning of this paragraph has, at least in certain cases, a negative answer and yields (in these cases) a considerable sharpening of Kaplansky's result on compact simple rings. A final settlement of the problem is in our opinion one of the most interesting questions still open in this area. For more details we refer to the comments concluding Chapter II.

Chapter III is concerned with equational compactness and topological compactness in the class of associative rings satisfying the ascending chain condition on left ideals. In contrast with the classes of rings studied in Chapter II, where the compact ones are always finite, this class of rings possesses infinite compact members, and therefore invites an investigation of the following more or less interrelated questions: Can the compact members carry more than one compact topology? Are the compact topologies occurring algebraically definable (e.g. as a radical topology, or as a product of discrete topologies on the factors of a product decomposition of the ring)? Can the compact rings be characterized algebraically? With regard to the uniqueness question we recall by way of analogy the old result of I. Gelfand [14] that a commutative semisimple algebra over the complex numbers admits only one topology converting it into a Banach algebra. Although there are compact groups admitting distinct compact topologies, we do not know of such examples in the class of associative rings (putting aside, of course, zero rings). We show in Chapter III that in the class of rings satisfying the ascending chain condition the compact members do have unique compact topologies and that these are always the (Jacobson) radical topologies (Theorems 6.8 and 6.9). These results extend the corresponding ones of Seth Warner [52] for (unital) noetherian rings. Crucial in the discussion, and of interest in its own right, is the fact that all the rings under scrutiny are compact as soon as they are equationally compact (Theorem 6.5). This chapter illustrates most clearly how well the algebraic property of equational compactness justifies itself in its rôle as an approximation of topological compactness, and how well the model theoretic input can be fruitfully applied. Certainly there is no apparent way to generalize the purely topological ring theoretic methods in [52] to obtain these results.

Chapter IV is a self-contained treatment of equational compactness, compactness, and injectivity in discriminator varieties, and represents an extension of (and incorporation of) the material in S. Bulman-Fleming & H. Werner [7] and that part of H. Werner [59] pertaining to injectivity in quasi-primal varieties. Our presentation avoids the use of the duality theory of K. Keimel and H. Werner [30], upon which the discussion in [7] is to a large extent based, although most of the elementary discriminator variety calculus used in [7] remains an essential tool in the deliberations. Without going into conceptual details at the moment, let it suffice to say that the results proved in this very much <u>universal</u> algebraic chapter apply to the class of "m-rings", that is, those rings satisfying an identity $x^m = x$, (alias "generalized" Boolean rings). In this chapter structural characterizations of equational compactness, compactness, and injectivity for members of such varieties are given (Theorems 8.16, 8.21, and 9.5). Moreover, the uniqueness-of-topology question already mentioned is answered positively for these classes. The compact topological members are iso- and homeomorphic to products of finite simple algebras (Theorem 8.21). With regard to the last statement, one is reminded of I. Kaplansky's structure theorem for compact topological semisimple rings - not surprising, as m-rings are semisimple. On the other hand the result is noteworthy in that it is universal algebraic in nature and fully independent of any specifically ring theoretic context. That the subject of injectivity occurs in Chapter IV, and only in Chapter IV, has a simple explanation: the only known equational classes of rings that we are aware of in which non-trivial injective objects may occur are equational classes of m-rings (the zero ring case aside). The ring theoretic precipitate of the characterization theorem for injective objects gives, for example, transparent conditions for an equationally compact m-ring to be injective in the equational class generated by it.

A compact topological (universal) algebra is, as observed at the outset, equationally compact, and so is every retract of an equationally compact algebra. (A retraction is a homomorphism onto a subalgebra leaving every element of the subalgebra fixed.) J. Mycielski asked in [38] whether every equationally compact algebra is in turn a retract of some compact algebra, a question naturally motivated by the fact, known since the 1950's, that the algebraically compact abelian groups are direct summands (= retracts) of compact abelian groups. Subsequently this question has been the subject of intense research; within various classes of universal algebras positive answers have been achieved, although the conjecture that it be true in general was proved false by W. Taylor [45] in 1969. Narrowing our attention to the class of associative rings, we shall see during the course of our discussion that a positive answer holds for all those classes of rings where a reasonalbe grasp on the equationally compact members can be achieved :- artinian rings, rings satisfying the ascending chain condition, and m-rings (not to mention those classes where the equationally compact members are finite). Thus the resolution of the Mycielski problem in the class of rings has a particular fascination - the class of rings displays most notably the strong interplay between equational compactness and topological compactness. Although we cannot give a definitive answer to the problem there remains however a most poignant motivation to terminate this exposition with a short discussion of the matter, namely the claim made by M.L. Kleĭner in an abstract from the conference on mathematical logic in Novosibirsk in 1974 that a counterexample does exist in the class of rings. All efforts on our part to check the correctness of his example as well as a lengthy correspondence with M.L. Kleĭner attempting to obtain his proof of the claim were to no avail.

The main result of Chapter V (and which at least on this side of the iron curtain must be considered new) is a further counterexample to the Mycielski Problem - an algebraic system possessing an abelian group as reduct. This demonstrates that there is at least nothing sacred about the abelian group (one such underlies every ring!) hindering the existence of a counterexample to the Mycielski Problem in the class of rings.

In conclusion let us stress once more that, although the results presented here are purely algebraic or topological algebraic in nature, the methods are in part model theoretic. As this same model theoretic potential may be lying dormant in connection with the investigation of analogous compactness questions in other classical areas, it seems worthwhile to make the "Brückenschlag" as transparent as possible in our presentation, with the hope that it might stimulate those researchers in the areas of topological groups, topological semigroups, etc., yet who are not intimate with the theory of equational compactness in universal algebras, to pose the question:

"What can equational compactness do for <u>me</u>?"

Should the compilation of our research in this form result in an encouragement for others to ask just this question, we feel the effort will have been justified by this fact alone.

CHAPTER I.
THE MODEL THEORETIC FRAMEWORK

Brevity and selfcontainedness are merits of a somewhat contradictory nature both of which however are to be striven for in an exposition of this format. We shall attempt to steer a course towards an acceptable resolution of this dilemma and present in this first chapter that background which we feel necessary to assure the ring theoretician not on easy terms with the elements of model theory confortable reading. Familiarity with the elements of point set topology, classical ring theory, abelian groups and topological rings, as given in [6], [11], [24], [26], [29], [35], [43], and [60], will be assumed. In our presentation of universal algebraic and model theoretic concepts we will restrict our brief account to that necessary to understand the results from the theory of equationally compact universal algebras (which we will only quote for later reference) on the one hand, and enough of the elementary properties of ultraproducts sufficient for following proofs using this fundamental construction (cf. Chapter IV) on the other. For a detailed account of all material in this introductory chapter (with the exception of the discussion centering around the structure topologies in §2) the reader may consult G. H. Wenzel's Habilitationsschrift [55]. Indeed, with the notable exceptions of [3], [4], [46] and [47], which have appeared since [55], the latter remains a comprehensive survey of the universal algebraic theory of equational compactness. Notation and terminology follow more or less that of G. Grätzer [16], and the reader is referred to that source for any unclarified universal algebraic concepts.

§1 Terminology and tools from the theory of models

A *type* is a quadruple $\tau = (F,n,R,m)$ where F and R are sets and n (resp. m) is a map from F (resp. R) into $\mathbb{N}_0$ (resp. $\mathbb{N}$). The elements of F are called "(fundamental) operation symbols" and those of R "(fundamental) relation symbols". For $f \in F$ $n(f)$ is the "arity" of the operation symbol f; similarly, $m(r)$ is the "arity" of $r \in R$. For a non-empty set A, a triple $\mathfrak{A} = \langle A; F^{\mathfrak{A}}, R^{\mathfrak{A}} \rangle$ is an *algebraic system of type* τ if $F^{\mathfrak{A}} = \{f^{\mathfrak{A}}; f \in F\}$ where for each $f \in F$, $f^{\mathfrak{A}}$ is an $n(f)$-ary operation on A, i.e., $f^{\mathfrak{A}} \in A^{A^{n(f)}}$, and similarly, if $R^{\mathfrak{A}}$ is the set of $r^{\mathfrak{A}}$'s where each $r^{\mathfrak{A}}$ is an $m(r)$-ary relation on A, i.e., $r^{\mathfrak{A}} \subseteq A^{m(r)}$, for $r \in R$. A is the *carrier set* of $\mathfrak{A}$. Note that a nullary operation symbol f just picks an element a from A, and we identify: $a = f^{\mathfrak{A}}$. The class of all algebraic systems of type τ denote by $\mathbb{K}(\tau)$. If R is empty (and in our deliberations this will most often be the case), then τ is an *algebra type* and the algebraic systems of type τ are (*universal*) *algebras*. In this case all symbolics pertaining to the (missing) relations will be suppressed. Also, *where the interpretation is clear from the context*, we will suppress the superscript "$\mathfrak{A}$", that is, we will notationally identify $F^{\mathfrak{A}}$ with F, $f^{\mathfrak{A}}$ with f, etc.; also, when clear from the context, the arity functions may be suppressed.

By way of illustration and also to fix terminology, the *class of* (*associative*) *rings* will be assumed to be that class of universal algebras of type $(\{+,-,\cdot,0\}, \{(+,2),(-,2),(\cdot,2),(0,0)\})$ satisfying the usual associative ring axioms.

A ring with identity is a ring possessing a unity element 1, whereas a unital ring will denote a ring with an identity considered as a universal algebra of an enriched type: the 1 is annexed as an additional nullary operation. (This subtle distinction is irrelevant when regarding equational compactness but is at least a priori of importance when considering categorical properties (cf. Chapter IV) of equational classes of rings.) If R is a ring with identity R_1 will denote the corresponding unital ring. (In all cases where confusion needn't arise, we will use the same Roman letter to denote the algebra as well as its underlying carrier set.)

Now fix $\mathfrak{A}, \mathfrak{B} \in K(\tau)$. $\mathfrak{A}$ is called a subsystem of $\mathfrak{B}$ (and $\mathfrak{B}$ an extension of $\mathfrak{A}$), denoted by $\mathfrak{A} \subseteq \mathfrak{B}$, if $A \subseteq B$ and for all $f \in F$ and $r \in R$, $f^{\mathfrak{A}} = f^{\mathfrak{B}}|_{A^{n(f)}}$ and $r^{\mathfrak{A}} = r^{\mathfrak{B}} \cap A^{m(r)}$; in short, the operations and relations on $\mathfrak{A}$ are obtained from those on $\mathfrak{B}$ by just restricting to A. A map $h : A \to B$ is a homomorphism from $\mathfrak{A}$ into $\mathfrak{B}$ (in symbols, $h : \mathfrak{A} \to \mathfrak{B}$) if for all $f \in F$ and all $a_1, \ldots, a_{n(f)} \in A$,

$$(1) \qquad h(f^{\mathfrak{A}}(a_1, \ldots, a_{n(f)})) = f^{\mathfrak{B}}(h(a_1), \ldots, h(a_{n(f)}))$$

and for all $r \in R$ and all $a_1, \ldots, a_{m(r)} \in A$,

$$(2) \qquad \text{if } r^{\mathfrak{A}}(a_1, \ldots, a_{m(r)}) \text{ then } r^{\mathfrak{B}}(h(a_1), \ldots, h(a_{m(r)})).$$

Using conditions (1) and (2) to define the operations and relations on h(A) the resulting algebraic system is called the homomorphic image of $\mathfrak{A}$ under h, denoted by $h(\mathfrak{A})$. If moreover h is injective or surjective then h is called a monomorphism or epimorphism, resp.; if $\mathfrak{A} = \mathfrak{B}$, then h is an endomorphism; if h is a monomorphism and $h(\mathfrak{A})$ is a subsystem of $\mathfrak{B}$, h is called an embedding (in case of algebras this definition is redundant, as every homomorphic image is a subalgebra of the range algebra); an epimorphic embedding is an isomorphism. A special kind of homomorphism is of particular importance: a homomorphism $h : \mathfrak{A} \to \mathfrak{B}$ is a retraction (of $\mathfrak{A}$ onto $\mathfrak{B}$) if there is an embedding $e : \mathfrak{B} \rightarrowtail \mathfrak{A}$ such that $h \circ e = id_B$ (the identity map on B), and we say that $\mathfrak{B}$ is a retract of $\mathfrak{A}$, if a retraction of $\mathfrak{A}$ onto $\mathfrak{B}$ exists. If $\mathbb{K}$ is a subclass of $\mathbb{K}(\tau)$, $\mathfrak{B}$ is called an absolute retract in $\mathbb{K}$ if $\mathfrak{B}$ is a retract of every $\mathfrak{A}$ in $\mathbb{K}$ which is an extension of $\mathfrak{B}$.

Finally, for a set $\mathfrak{A}_i$, $i \in I$, of algebraic systems of type τ the product $\mathfrak{A} = \prod(\mathfrak{A}_i ; i \in I)$ is defined as follows: the carrier set of $\mathfrak{A}$ is the cartesian product $\prod(A_i ; i \in I)$ and operations and relations are defined componentwise:

$$f^{\mathfrak{A}}(a_1, \cdots, a_{n(f)})(i) = f^{\mathfrak{A}_i}(a_1(i), \cdots, a_{n(f)}(i))$$

and

$$r^{\mathfrak{A}}(a_1, \cdots, a_{m(r)}) \text{ iff } r^{\mathfrak{A}_i}(a_1(i), \cdots, a_{m(r)}(i)) \text{ for all } i \in I.$$

Given the above, $\mathfrak{B}$ is a subdirect product of the $\mathfrak{A}_i$, $i \in I$, if $\mathfrak{B}$ is a subsystem of $\mathfrak{A}$, and $p_{i_o}(B) = A_{i_o}$ for each of the canonical projections $p_{i_o} : \prod(A_i ; i \in I) \twoheadrightarrow A_{i_o}$.

For the scope of our investigations we require a slight broadening of the usual concept of "polynomial": for an ordinal α let X_α denote the following set of "variables", well-ordered by the ordinal α: $X_\alpha = \{x_\gamma ; \gamma < \alpha\}$. For our fixed type τ let $P_\alpha(\tau)$ denote the set of <u>α-ary polynomial symbols of type τ</u>, inductively defined by the rules

(1) $X_\alpha \cup \{f \in F;\ n(f) = 0\} \subseteq P_\alpha(\tau)$, and

(2) if $p_1,\cdot\cdot,p_{n(f)} \in P_\alpha(\tau)$, $f \in F$, then $f(p_1,\cdot\cdot,p_{n(f)}) \in P_\alpha(\tau)$.

Similarly, let the <u>α-ary relational predicate symbols of type τ</u> be given by

$$Q_\alpha(\tau) = \{r(y_1,\cdots,y_{m(r)});\ r \in R,\ y_1,\cdots,y_{m(r)} \in X_\alpha\}.$$

For $\mathfrak{A} \in \mathbb{K}(\tau)$ let $P_\alpha(\mathfrak{A})$ denote the set of <u>α-ary polynomials on $\mathfrak{A}$</u>, $P_\alpha(\mathfrak{A}) = \{p^{\mathfrak{A}};\ p \in P_\alpha(\tau)\}$, where $p^{\mathfrak{A}} : A^\alpha \rightarrow A$ is defined inductively by

(1) $x_\gamma(a) = a(\gamma)$ and $f(a) = f^{\mathfrak{A}}$ $(a \in A^\alpha, \gamma < \alpha, n(f) = 0)$, and

(2) $p^{\mathfrak{A}}(a) = f^{\mathfrak{A}}(p_1^{\mathfrak{A}}(a),\cdot\cdot,p_{n(f)}^{\mathfrak{A}}(a))$ if $p = f(p_1,\cdot\cdot,p_{n(f)})$.

Proceding, let $Q_\alpha(\mathfrak{A})$ denote the set of <u>α-ary relational predicates on $\mathfrak{A}$</u> defined by

$$r\ (y_1,\cdots,y_{m(r)})(a) \text{ iff } r\ (y_1(a),\cdots,y_{m(r)}(a)),$$

for $a \in A^\alpha$ and $y_1,\cdots,y_{m(r)} \in X_\alpha$. Via an obvious modification of the above definition we define, for a subset C of A, the set $P_\alpha^C(\mathfrak{A})$ (resp. $Q_\alpha^C(\mathfrak{A})$) of <u>α-ary polynomials (on $\mathfrak{A}$)</u> (resp. <u>relational predicates (on $\mathfrak{A}$)) with constants in C</u>, by allowing, in the inducing polynomial and relational predicate symbols, some of the occurring variables to be replaced by elements of C. Note that $P_\alpha^\emptyset(\mathfrak{A}) = P_\alpha(\mathfrak{A})$. The cartesian product $P_\alpha^C(\mathfrak{A}) \times P_\alpha^C(\mathfrak{A})$ is the set of <u>α-ary polynomial equations (on $\mathfrak{A}$) with constants in C</u>, and the union of this set with $Q_\alpha^C(\mathfrak{A})$ is the set <u>$At_\alpha^C(\mathfrak{A})$</u> of <u>$\alpha$-ary atomic formulas (on $\mathfrak{A}$) with constants in C</u>.

Before pursuing the model theoretic build-up further, we pause to consider the special case where τ is an algebra type and concentrate on the set $P_{\omega_0}(\tau)\times P_{\omega_0}(\tau)$, also called the <u>polynomial identities of type τ</u>. If $\mathfrak{A}\in\mathbb{K}(\tau)$, we say that $\mathfrak{A}$ <u>satisfies the polynomial identity (p,q)</u> (often written more suggestively as "p=q"), if $p^{\mathfrak{A}} = q^{\mathfrak{A}}$; we then call $\mathfrak{A}$ a <u>model of p=q</u>. A subclass $\mathbb{K}$ of $\mathbb{K}(\tau)$ is an <u>equational class</u> (or <u>variety</u>) if there is a set Σ of polynomial identities of type τ such that K is the subclass of $K(\tau)$ consisting of precisely those algebras satisfying each identity in Σ ; we write: $K = \mathrm{Mod}(\Sigma)$, the <u>model class of Σ</u>. Numerous "classical" classes of algebras :- groups, abelian groups, rings, modules, (classical) algebras, lattices, Boolean algebras, etc. etc. - are equational classes simply because the defining systems of "axioms" can be interpreted, in these cases, as polynomial identities as soon as the type is chosen rich enough to include the operations which occur in the axioms. A fundamental theorem of G. Birkhoff says that the equational classes are precisely those classes satisfying $\mathbb{K} = \mathbb{HSP}(\mathbb{K})$, that is, those classes closed under the formation of direct products, subalgebras, and homomorphic images (cf. "recurring notations" p. 9); moreover, for any class $\mathbb{K}$ in $\mathbb{K}(\tau)$ $\mathbb{HSP}(\mathbb{K})$ is an equational class (and therefore the smallest one) containing $\mathbb{K}$; we write sometimes $\mathbb{V}(\mathbb{K})$ for $\mathbb{HSP}(\mathbb{K})$ and refer to the "variety generated by $\mathbb{K}$".

When reference is made to a categorical property of an (equational) class, the class is tacitly assumed to have the following categorical interpretation: objects are the algebras and morphisms are the homomorphisms; note that we have no "empty homomorphisms" as in A. Day [10].

Familiarity with the concept of a congruence relation on an algebra, i.e., those relations occurring as kernels of homomorphisms emanating from it, as well as the so-called Homomorphism and Isomorphism Theorems will be assumed, as these are the natural (and trivial) generalizations of their analogues in the classical group and ring theoretical context. The principal congruence generated by $a,b \in A$, namely the unique smallest congruence on $\mathfrak{A}$ identifying a and b, is denoted by $\Theta(a,b)$ (we will of course write $(a-b)$ in all specifically ring theoretic contexts). Recall that the set $C(\mathfrak{A})$ of all congruences on an algebra $\mathfrak{A}$ is a complete sublattice, denoted by $\mathcal{L}(\mathfrak{A})$, of the lattice of all equivalence relations on A; moreover $\mathcal{L}(\mathfrak{A})$ is algebraic, namely every element is the (infinite) join of compact elements (x is <u>compact</u> if, whenever $x = \bigvee(x_i;i\in I)$ then $x = x_{i_1}\vee\cdots\vee x_{i_n}$ for suitable $i_1,\cdots i_n \in I$). In $\mathcal{L}(\mathfrak{A})$ the finite joins of principal congruences are precisely the compact elements (cf. [16, Ch. 2]). Finally, $\mathfrak{A}$ is <u>subdirectly irreducible</u> if, whenever $\mathfrak{A}$ is isomorphic to a subdirect product of a family $\mathfrak{A}_i$, $i \in I$, then $\mathfrak{A}$ is already isomorphic via the canonical projection to one of the factors. Equivalently, $\mathcal{L}(\mathfrak{A})$ is atomic with exactly one atom. Another basic theorem due to G. Birkhoff states that every algebra $\mathfrak{A}$ is isomorphic to a subdirect product of subdirectly irreducible algebras, hence the interest displayed for the subdirectly irreducibles in structure theoretic considerations.

Let us pursue now the construction of our language of

(first order) elementary formulas and begin by blowing some semantic life into the set $At_{\alpha}^{C}(\mathfrak{A})$ of α-ary atomic formulas on $\mathfrak{A}$ with constants in C, where $\mathfrak{A}$ is an algebraic system of some arbitrary type τ: For $\phi \in At_{\alpha}^{C}(\mathfrak{A})$ and a subsystem $\mathfrak{B}$ of $\mathfrak{A}$ with $C \subseteq B$ we define $b \in B^{\alpha}$ to be a <u>solution of ϕ in $\mathfrak{B}$</u> if

(1) $p^{\mathfrak{A}}(b) = q^{\mathfrak{A}}(b)$ in case $\phi = (p^{\mathfrak{A}}, q^{\mathfrak{A}})$, or

(2) $\phi(b)$ in case $\phi \in Q_{\alpha}^{C}(\mathfrak{A})$.

Denote by $Sol_{\mathfrak{B}}(\phi)$ the subset of B^{α} of all solutions of ϕ in $\mathfrak{B}$. A set of α-ary (atomic) formulas $\{\phi_i;\ i \in I\}$ is <u>solvable in $\mathfrak{B}$</u> in case there is a $b \in B^{\alpha}$ which is a common solution in $\mathfrak{B}$ of each ϕ_i, $i \in I$. A subsystem $\mathfrak{B}$ of $\mathfrak{A}$ is a <u>pure subsystem of $\mathfrak{A}$</u> if every finite subset of $At_{\omega_0}^{B}(\mathfrak{A})$ which is solvable in $\mathfrak{A}$ is solvable in $\mathfrak{B}$; an embedding $h : \mathfrak{B} \to \mathfrak{A}$ is <u>pure</u> if $h(\mathfrak{B})$ is a pure subsystem of $\mathfrak{A}$. For example, if G is a pure subgroup of the abelian group H, then it follows that $n \cdot H \cap G = n \cdot G$ for all $n \in \mathbb{N}$, because the equation $n \cdot x = g$ $(g \in G)$ must be solvable in G if it is solvable in H, that is, G is a "pure" subgroup of H in the group theoretic sense (see e.g. I. Kaplansky [29]). S. Gacsâlyi [13a] showed that the converse is also true: group "purity" implies, for groups, the above defined purity; this proved to be a keystone for the theory of equationally compact abelian groups as developed by S. Balcerzyk [1], I. Kaplansky [29], and J. Loś [34] in the 1950's, and this development in turn served J. Mycielski as a concrete motivation for the introduction and initial investigation of equational compactness in the general algebraic context.

Now let τ and τ' be two (arbitrary) types and suppose that $\mathfrak{A} = \langle A; F, R\rangle \in \mathbb{K}(\tau)$ and $\mathfrak{A}' = \langle A'; F', R'\rangle \in \mathbb{K}(\tau')$. We say that $\mathfrak{A}$ is a <u>reduct</u> of $\mathfrak{A}'$ if, first, $A = A'$ and second, for every ordinal α and every $\phi \in At_\alpha^A(\mathfrak{A})$ there is a $\phi' \in At_\alpha^A(\mathfrak{A}')$ such that $Sol_{\mathfrak{A}}(\phi) = Sol_{\mathfrak{A}'}(\phi')$. Loosely spoken, $\mathfrak{A}$ is a reduct of $\mathfrak{A}'$ if, on the set A, the algebraic structure imposed by τ' is at least as rich as that imposed by τ. $\mathfrak{A}$ is a <u>strong reduct</u> of $\mathfrak{A}'$ if, first, $A = A'$ and second, every fundamental operation on $\mathfrak{A}$ is a polynomial on $\mathfrak{A}'$ and every fundamental relation on $\mathfrak{A}$ is a fundamental relation on $\mathfrak{A}'$. Clearly strong reducts are reducts. Call $\mathfrak{A}$ a (<u>strong</u>) <u>subreduct</u> of $\mathfrak{A}'$ if $\mathfrak{A}$ is a subsystem of a (strong) reduct of $\mathfrak{A}'$, and call $\mathfrak{A}$ and $\mathfrak{A}'$ (<u>strong</u>) <u>reduct equivalent</u> if each is a (strong) reduct of the other.

Let us illustrate and again fix some terminology by the way. If R is a ring then R possesses as strong reduct its underlying abelian group, which we denote by R^+. If G is an abelian group then the ring G_o defined by $(G_o)^+ = G$ and letting all products be zero is strongly reduct equivalent to G, and is called the <u>zero ring over G</u>. If $B = \langle B; \wedge, \vee, 0, 1, '\rangle$ is a Boolean algebra, then the unital ring $B_r = \langle B; +, -, \cdot, 0, 1\rangle$ defined by $a+b = (a\vee b)\wedge(a\wedge b)'$, $-a = a'$ and $a\cdot b = a\wedge b$ is the <u>Boolean ring associated with B</u>; B and B_r are strong reduct equivalent, as the operations of B are easily regained as polynomials in the ring B_r. All the central concepts investigated in this exposition are invariants with respect to strong reduct equivalence, and most with respect to reduct equivalence, and therefore where the context allows a distinction between (strong) reduct equivalents will not be made.

The set $L_\alpha^C(\mathfrak{A})$ of __α-ary first order formulas on $\mathfrak{A}$ with constants in C__ is now defined inductively, using $At_\alpha^C(\mathfrak{A})$ as building blocks, as follows:

(1) $At_\alpha^C(\mathfrak{A}) \subseteq L_\alpha^C(\mathfrak{A})$.

(2) If ϕ_1 and ϕ_2 are in $L_\alpha^C(\mathfrak{A})$, then $(\phi_1 \wedge \phi_2)$ and $(\phi_1 \vee \phi_2)$ are in $L_\alpha^C(\mathfrak{A})$ (the "conjunction" resp. "disjunction" of ϕ_1 and ϕ_2).

(3) If ϕ is in $L_\alpha^C(\mathfrak{A})$ then for all $0 \leq \beta < \alpha$ $(\exists x_\beta)\phi$ $(\forall x_\beta)\phi$ are in $L_\alpha^C(\mathfrak{A})$ (the "existential" resp. "universal quantification of the variable x_β in ϕ"). In both cases the formula ϕ is called the "scope" of $(\exists x_\beta)$ resp. $(\forall x_\beta)$.

(4) If ϕ is in $L_\alpha^C(\mathfrak{A})$ then $\neg\phi$ is in $L^C(\mathfrak{A})$ (the "negation" of ϕ).

Let $L^C(\mathfrak{A})$ denote the class of __first order formulas on $\mathfrak{A}$ with constants in C__ :

$$L^C(\mathfrak{A}) = \bigcup(L_\alpha^C(\mathfrak{A}) \; ; \; \alpha \text{ ordinal}).$$

Similarly, $At^C(\mathfrak{A})$ will denote the class of all atomic formulas. In all contexts where $C = \emptyset$ the symbol will be suppressed. Members of $L^A(\mathfrak{A})$ we will often refer to as (first order) formulas "over $\mathfrak{A}$". A (first order) formula is __positive__ if it is constructed using only the rules (1) - (3) above. For $\phi \in L_\alpha^C(\mathfrak{A})$ define the set $Sol_{\mathfrak{B}}(\phi)$ of __solutions of ϕ in $\mathfrak{B}$__ (where $\mathfrak{B}$ is a subsystem of $\mathfrak{A}$ containing C) inductively as follows:

(1) For $\phi \in At^C_\alpha(\mathfrak{A})$, $Sol_{\mathfrak{B}}(\phi)$ has already been defined.

(2) $Sol_{\mathfrak{B}}((\phi_1 \wedge \phi_2)) = Sol_{\mathfrak{B}}(\phi_1) \cap Sol_{\mathfrak{B}}(\phi_2)$,

$Sol_{\mathfrak{B}}((\phi_1 \vee \phi_2)) = Sol_{\mathfrak{B}}(\phi_1) \cup Sol_{\mathfrak{B}}(\phi_2)$.

(3) For $b \in B^\alpha$, $a \in B$, and $0 \leq \beta < \alpha$, define $b(\beta|a) \in B^\alpha$ by

$b(\beta|a)(\delta) = b(\delta)$ if $\delta \neq \beta$, and $b(\beta|a)(\beta) = a$.

Then

$Sol_{\mathfrak{B}}((\exists x_\beta)\phi) = \{b \in B^\alpha;\ b(\beta|a) \in Sol_{\mathfrak{B}}(\phi)$ for some $a \in B\}$,

$Sol_{\mathfrak{B}}((\forall x_\beta)\phi) = \{b \in B^\alpha;\ b(\beta|a) \in Sol_{\mathfrak{B}}(\phi)$ for all $a \in B\}$.

(4) $Sol_{\mathfrak{B}}(\neg\phi) = B^\alpha \setminus Sol_{\mathfrak{B}}(\phi)$.

As was the case with subsets of $At^C_\alpha(\mathfrak{A})$, we say that a subset $\Sigma = \{\phi_i;\ i \in I\}$ of $L^C_\alpha(\mathfrak{A})$ is <u>solvable in</u> $\mathfrak{B}$ provided there is a $b \in B^\alpha$ which is a common solution of each ϕ_i, $i \in I$. Σ is <u>finitely solvable in</u> $\mathfrak{B}$ if each finite subset of Σ is solvable in $\mathfrak{B}$. Obviously the definition of solvability is independent of the choice of α as soon as α is chosen large enough for $L^C_\alpha(\mathfrak{A})$ to contain Σ . Moreover, as any sub<u>set</u> of the class $L^C(\mathfrak{A})$ is contained in some $L^C_\alpha(\mathfrak{A})$, it follows that "solvability in $\mathfrak{B}$ of subsets of the <u>class</u> $L^C(\mathfrak{A})$" is a well-defined notion. Two α-ary formulas over $\mathfrak{A}$ are <u>equivalent in</u> $\mathfrak{A}$ if their solution sets coincide, and again this notion does not depend on (sufficiently large) α.

If (Qx_β) occurs in the formula ϕ , where Q is one of $\exists$ or $\forall$, then we say that the occurrence of x_β in (Qx_β) as well as any occurrence of x_β in the scope of (Qx_β) is <u>bound in</u> ϕ ; a variable is <u>free in</u> ϕ if it has a non-bound occurrence in ϕ . A <u>sentence</u> is a formula with no free variables, and it is clear that a sentence ϕ in $L^C_\alpha(\mathfrak{A})$ either <u>holds in</u> $\mathfrak{A}$ or <u>does not hold in</u> $\mathfrak{A}$ depending on whether $Sol_{\mathfrak{A}}(\phi) = A^\alpha$ or $Sol_{\mathfrak{A}}(\phi) = \emptyset$. By way of illustration the formula $(\forall x_0)(\exists x_1)(x_0 = 0 \vee x_0 \cdot x_1 = 1)$ is a sentence, ϕ, over every unital ring, and for any commutative unital ring R ϕ will hold in R precisely when R is a field.

Finally, a subsystem $\mathfrak{B}$ of an algebraic system $\mathfrak{A}$ is <u>an elementary subsystem of</u> $\mathfrak{A}$ (and $\mathfrak{A}$ is an <u>elementary extension of</u> $\mathfrak{B}$) if, for each $\phi \in L_{\omega_0}(\mathfrak{A})$, $Sol_{\mathfrak{B}}(\phi) = B^{\omega_0} \cap Sol_{\mathfrak{A}}(\phi)$, that is, an ω_0-tuple of elements from B is a solution of ϕ in $\mathfrak{B}$ if and only if it is a solution of ϕ in $\mathfrak{A}$. Again, we call an embedding $h : \mathfrak{B} \to \mathfrak{A}$ <u>elementary</u> if $h(\mathfrak{B})$ is an elementary subsystem of $\mathfrak{A}$. A bit of reflection soon reveals what a strong condition this is; e.g., note that an elementary subsystem is a pure subsystem and, indeed, much more in general. For example, a finite algebraic system possesses no proper elementary extension.

In the preceding discussion we have, in one sense, put the cart before the horse in our construction of first order formulas, by defining a formula depending on a pregiven algebraic system $\mathfrak{A}$. That is simply a reflection of the totally semantical bent of this exposition, and we are convinced that, having seen and grasped the "cart", the reader will find in retrospect no difficulty in grasping the "horse". What is meant is the following: Define, for a fixed type τ, the class of <u>first order formulas of type</u> τ, $L(\tau)$ by beginning - in step (1) of the construction of α-ary formulas on $\mathfrak{A}$ with constants in C - with the <u>α-ary atomic formulas of type</u> τ, $At_\alpha(\tau) = P_\alpha(\tau) \times P_\alpha(\tau) \cup Q_\alpha(\tau)$, instead of with $At_\alpha^C(\mathfrak{A})$. $L(\tau)$ is thus a class of formal sequences of symbols with a priori no inherent notion of "solvability". On the other hand, given <u>any</u> member $\mathfrak{A}$ of $\mathbb{K}(\tau)$, a formula $\phi \in L(\tau)$ induces uniquely a formula in $L(\mathfrak{A})$ by affixing the superscript "$\mathfrak{A}$" to all occurring operation and relation symbols. In this sense we can interpret in a canonical fashion any $\phi \in L(\tau)$ in <u>any</u> algebraic system of type τ and speak of "solutions of ϕ in $\mathfrak{A}$", "solutions of ϕ in $\mathfrak{B}$", etc. by regarding the <u>induced</u> formula, once on $\mathfrak{A}$, the other time on $\mathfrak{B}$.

This concludes our rather tedious expedition through the forest of basic notation and terminology, and we are now in a position to put some coherence - and justification - into all of this beginning with that notion, the discovery and subsequent investigation of which represents one of the major contributions to modern-day model theory :- the ultraproduct. To pursue this aim recall first that for a set I, a <u>filter on I</u> is a sublattice of the power set lattice 2^I closed under the formation of supersets (i.e., a dual ideal of 2^I). A filter F on I is an <u>ultrafilter</u> if it is maximal in the family of all proper filters on I. E.g., the <u>principal filter</u> generated by a singleton - $[i) = \{X \subseteq I;\ i \in X\}$ - is an ultrafilter. It is well known (and easily verified) that a filter F on I is maximal iff it is <u>prime</u> (i.e., $X \cup Y \in F$ iff $X \in F$ or $Y \in F$) and this in turn holds iff F satisfies the condition: $X \in F$ iff $I \setminus X \notin F$, for all $X \subseteq I$. This last condition simply says that ultrafilters on I are the $0,1$ - valued measures on I.

Now let $\{\mathfrak{A}_i\ ;\ i \in I\}$ be a family of algebraic systems in $K(\tau)$, and let F be an ultrafilter on I. Define a relation, θ_F, on the product $\mathfrak{A} = \prod(\mathfrak{A}_i\ ;\ i \in I)$ as follows: for $a,b \in A$ let

$$a \equiv b\ (\theta_F) \quad \text{iff} \quad \{i \in I\ ;\ a(i) = b(i)\} \in F\ .$$

In other words I-tuples are identified if they agree on a "large" subset of I ("largeness" meaning belonging to F). It follows immediately from the filter properties first, that θ_F is an equivalence relation on $A = \prod(A_i; i \in I)$ and, second, that the following declarations on the quotient set A/θ_F are well-defined for a fundamental operation symbol f and fundamental relation symbol r :

(1) $f^{A/\theta_F}(\hat{a}_1,\dots,\hat{a}_{n(f)}) := \widehat{f^{\mathfrak{A}}(a_1,\dots,a_{n(f)})}$

(2) $r^{A/\theta_F}(\hat{a}_1,\dots,\hat{a}_{m(r)})$:iff $\{i \in I;\ r^{\mathfrak{A}_i}(a_1(i),\dots,a_{m(r)}(i))\} \in F$

where for $a \in A$ $\hat{a}$ stands for the equivalence block $[a]\theta_F$. The above defined operations and relations turn A/θ_F into an algebraic system of type τ, called the reduced product of the family $\{\mathfrak{A}_i; i \in I\}$ over F, and is denoted by $\prod_F(\mathfrak{A}_i; i \in I)$. If F is an ultrafilter - and this will be the case of primary interest to us - $\prod_F(\mathfrak{A}_i; i \in I)$ is the ultraproduct of $\{\mathfrak{A}_i; i \in I\}$ over F. If all the $\mathfrak{A}_i$'s are identical, equal say to the algebraic system $\mathfrak{B}$, we speak of the reduced power (resp. ultrapower) of $\mathfrak{B}$ over F, and write $\mathfrak{B}^I_F$. Note that the map $a \mapsto \hat{a}$ is a homomorphism of $\mathfrak{A} = \prod(\mathfrak{A}_i; i \in I)$ onto $\prod_F(\mathfrak{A}_i; i \in I)$ with kernel θ_F. If τ is an algebra type θ_F is therefore a congruence on $\prod(\mathfrak{A}_i; i \in I)$, the congruence induced by F. We call an arbitrary congruence θ on $\mathfrak{A}$ an (ultra-)filter congruence if $\theta = \theta_F$ for some (ultra-)filter F. The rôle played by filter congruences in congruence distributive varieties, as exemplified in particular by "Jónsson's Lemma", will be amply illustrated in the investigations in Chapter IV.

For a filter F on I and arbitrary $M \in F$ denote by F_M the filter $\{M \cap X;\ X \in F\}$ on M, the restriction of F to M.

A very easy but useful fact is that the canonical projection of $\prod(\mathfrak{A}_i;\ i \in I)$ onto $\prod(\mathfrak{A}_i;\ i \in M)$ induces an isomorphism between $\prod_F(\mathfrak{A}_i;\ i \in I)$ and $\prod_{F_M}(\mathfrak{A}_i;\ i \in M)$. In particular, $\prod_F(\mathfrak{A}_i;\ i \in I) \cong \mathfrak{A}_{i_o}$ if F is the principal filter $[i_o)$.

A close look at the relations (1) and (2) defining the structure of $\prod_F(\mathfrak{A}_i;\ i \in I)$ yields the following observation: For any ω_o-tuple $\bar{a} = (a_o, a_1, \cdots)$ of elements from $A = \prod(A_i;\ i \in I)$ denote by $\hat{a}$ the ω_o-tuple $(\hat{a}_o, \hat{a}_1, \cdots)$ of elements from A/θ_F, and by $\bar{a}(i)$ the ω_o-tuple $(a_o(i), a_1(i), \cdots) \in A_i^{\omega_o}$. Then for any $\phi \in At_{\omega_o}(\tau)$, $\hat{a}$ is a solution of ϕ in $\prod_F(\mathfrak{A}_i;\ i \in I)$ if, and only if,

$$\{i \in I\ ;\ \bar{a}(i) \text{ is a solution of } \phi \text{ in } \mathfrak{A}_i\} \in F\ .$$

It is precisely the generalization of this simple observation to <u>arbitrary</u> first order formulas where the ultraproduct construction has its central model theoretic significance. This is the important

<u>1.1. Proposition (J. Loś [33])</u> Let $\prod_F(\mathfrak{A}_i;\ i \in I)$ be an ultraproduct and let $\phi \in L_{\omega_o}(\tau)$. Then for $\bar{a} \in \prod(A_i;\ i \in I)^{\omega_o}$,

$$\hat{a} \in Sol_{\prod_F(\mathfrak{A}_i;\ i \in I)}(\phi) \quad \text{iff} \quad \{i \in I\ ;\ \bar{a}(i) \in Sol_{\mathfrak{A}_i}(\phi)\} \in F\ .$$

In short, satisfiability in $\prod_F(\mathfrak{A}_i;\ i \in I)$ is componentwise satisfiability on a large subset of the $\mathfrak{A}_i$'s. For example, if a large number of the $\mathfrak{A}_i$'s have n-element carrier sets, then $\prod_F(\mathfrak{A}_i;\ i \in I)$ has exactly n elements (and vica versa), because "X has n elements" is characterized by a first order sentence on X . For another example, observe that an ultraproduct of fields is a field, as the field axioms are first order sentences.

For A and I non-empty sets and $a \in A$ denote by $(a)_I$ the "diagonal" element of the power set A^I defined by $(a)_I(i) = a$, $i \in I$. An important consequence of Proposition 1 is the following

1.2. Proposition Let $\mathfrak{A}^I_F$ be an ultrapower of the algebraic system $\mathfrak{A}$. Then

$$d : \mathfrak{A} \to \mathfrak{A}^I_F$$
$$a \mapsto \widehat{(a)_I}$$

is an elementary embedding of $\mathfrak{A}$ into $\mathfrak{A}^I_F$, called the diagonal, or canonical embedding.

A comprehensive development of the theory of reduced products was first undertaken - seven years after the cited paper of J. Loś - by T. Frayne, A.C. Morel and D.S. Scott in [13]. The properties given by Propositions 1 and 2 will however suffice for the applications given in this treatise, and, indeed, the results from the theory of equational compactness which will be presented without proof in the next section rely essentially, in their model theoretic aspects, on these results alone.

§2 From the theory of equationally compact universal algebras; the structure topologies

The purpose of this section is two-fold. First we will present those results on equational compactness which form the general algebraic foundation of the (for the large part) ring theoretic investigations of the subsequent chapters. As a detailed, unified, and indeed substantially generalized account of this theory is given in G. H. Wenzel [55], we shall remain true to the style of §1 and restrict our account to brief statements of the results, pointing out their interrelations and the model theoretic input, however without giving proofs. Our second goal will be to define the "structure topology" and derive those basic properties that will serve in good stead as tools throughout this thesis. The structure topology provides, very simply, a topological interpretation of equational compactness; it has the desirable property of algebraic invariance together with a good portion of "topological-algebraic" behaviour, but has at the same time the disadvantage of not turning in general an algebra into a (compact) topological algebra in any canonical way. Indeed, the "poorness" of the structure topology serves in a sense as a barometer of how "far apart" equational and topological compactness in a particular class of algebras are.

Putting heuristics aside, let us begin with the definition of that concept central to all the deliberations to follow.

2.1. Definition (J. Mycielski [38]). Let $\mathfrak{A}$ be an algebraic system and let L be a subclass of $L^A(\mathfrak{A})$. $\mathfrak{A}$ is said to be

L - compact if every subset of L which is finitely solvable in $\mathfrak{A}$ is solvable in $\mathfrak{A}$ (for which we shall also synonomously say, $\mathfrak{A}$ has the relative solvability property on subsets of L). Thus if L is the class of all atomic formulas over $\mathfrak{A}$, all positive formulas over $\mathfrak{A}$, ... , we say then that $\mathfrak{A}$ is atomic compact, positively compact, ... If $\mathfrak{A}$ is an algebra, then $At^A(\mathfrak{A})$ is the class of all polynomial equations over $\mathfrak{A}$, and atomic compactness is then generally called equational compactness.

Before proceding with the model theoretic development of these concepts, let us pause to consider a fundamental connection between equational compactness and topological algebra as such. It is this feature which has made the theory of equational compactness of utmost relevance to classical areas within topological algebra, and it is precisely this feature which is a recurring Leitmotiv of this exposition.

If $\mathfrak{A}$ is an algebraic system of type $\tau = (F,n,R,m)$ and $\mathcal{T}$ is a Hausdorff topology on A we call $(\mathfrak{A},\mathcal{T})$ a topological algebraic system (and $\mathcal{T}$ a topology for $\mathfrak{A}$) if $\mathcal{T}$ is compatible with the structure of $\mathfrak{A}$, that is, $f : A^{n(f)} \to A$ is continuous for all $f \in F$, and r is a closed subset of $A^{m(r)}$ for all $r \in R$ (products of A being endowed with their product topologies). Call $(\mathfrak{A},\mathcal{T})$ a compact topological algebraic system if, in addition, $\mathcal{T}$ is compact. A algebraic system $\mathfrak{A}$ is compact if there exists a compact topology for $\mathfrak{A}$. ("compact" will always mean quasi-compact and Hausdorff.) Now if $(\mathfrak{A},\mathcal{T})$ is a compact topological algebraic system, then for any α-ary atomic formula ϕ

over $\mathfrak{A}$, $Sol_{\mathfrak{A}}(\phi)$ is a closed aubset of A^{α} endowed with the compact Tychonoff product topology; hence the atomic compactness of $\mathfrak{A}$ is implied by the compactness of the topological spaces A^{α} . Thus compact algebraic systems are atomic compact. An even more elementary fact is that, if $\mathfrak{B}$ is a retract of the atomic compact algebraic system $\mathfrak{A}$, then $\mathfrak{B}$ is atomic compact, as a solution in $\mathfrak{A}$ of any atomic formula ϕ over $\mathfrak{B}$ is mapped by the retraction onto a solution <u>in $\mathfrak{B}$</u> of ϕ . Putting this together we have the following fundamental observation :

<u>2.2. Proposition (J. Mycielski [38])</u>. A retract of a compact algebraic system is atomic compact.

It was known already at the time (1964) that the equationally compact abelian groups were <u>precisely</u> the retracts (i.e., direct summands) of the compact ones, and thus the question as to the validity of the converse of Proposition 2 for arbitrary algebras was, naturally enough, raised. In spite of the fact that the converse has since been negated by counterexamples (due to W. Taylor) the investigation of the question in <u>specific</u> classes of algebras still remains a subject of intense interest and continued research. As we will be dealing with this question in various classes of rings, let us shortly formulate the problem : If $\mathbb{K}$ is a class of algebraic systems, the <u>Mycielski question for $\mathbb{K}$</u> is the following - "Is every atomic compact member of $\mathbb{K}$ retract of some compact algebraic system?"

In a series of papers which appeared during the middle 1960's J. Mycielski, C. Ryll-Nardzewski, and B. Weglorz initiated the development of the theory of equational compactness (see

[38],[39],[53],[54]). The following characterization theorem due to B. Weglorz is of basic importance, not only for the general theory but also for our specific ring theoretic investigations.

<u>2.3. Proposition (B. Weglorz [53])</u>. For an algebraic system the following are equivalent:

(i) $\mathfrak{A}$ is positively compact

(ii) $\mathfrak{A}$ is atomic compact

(iii) $\mathfrak{A}$ is a retract of every pure extension

(iv) $\mathfrak{A}$ is a retract of every elementary extension

(v) $\mathfrak{A}$ is a retract of every ultrapower

Condition (iii) in the above may be replaced by (the a priori stronger)

(iii)' $\mathfrak{A}$ is pure-injective

($\mathfrak{A}$ is <u>pure-injective</u> if every homomorphism $h : \mathfrak{A} \to \mathfrak{B}$ can be extended to any pure extension of $\mathfrak{A}$. For a proof of (ii)$\Leftrightarrow$(iii)' consult [55]. Condition (iii)' illuminates the relative position of atomic compactness vis à vis injectivity; cf. §9.)

In passing, a brief comment on some of the ingredients of the proof of Proposition 3 - (iv) $\Rightarrow$ (v) is given by Proposition 1.2 (enter Loś' Theorem!). Loś' Theorem is also essential for the implication (v)$\Rightarrow$(i). Here an argument analogous to the proof using ultraproducts of the famed Compactness Theorem (see e.g. [16]) is used: for a finitely solvable set Σ of first order formulas over $\mathfrak{A}$ there exists an ultrafilter F on the set I of finite subsets of Σ such that $\mathfrak{A}_F^I$ possesses a solution of Σ. In case the formulas are all positive, that solution is mapped into a solution in $\mathfrak{A}$ by the retraction guaranteed by (v).

It is the equivalence of (i) and (ii) that will have particular resonance in our considerations. Just to illustrate its power, observe that the Alexander Subbase Theorem in topology falls out as an example: namely, if A is a topological space and R is a subbase of closed sets, then, viewing members of R as unary relations on A, $\mathfrak{A} = \langle A; R\rangle$ is an algebraic system. If R has the "compactness property", i.e., if every family in R with the finite intersection property has a non-empty intersection, then that just means that $\mathfrak{A}$ is atomic compact; thus $\mathfrak{A}$ is positively compact so, in particular, (concentrating on the class of disjunctions of atomic formulas) the base of closed sets generated by R also has the compactness property; that is the statement of the Alexander Subbase Theorem.

For the topological interpretation of atomic compactness a further characterization of atomic compactness is essential.

<u>2.4. Proposition (J. Mycielski & C. Ryll-Nardzewski [39])</u>. An algebraic system $\mathfrak{A}$ is atomic compact iff $\mathfrak{A}$ has the relative solvability property on sets of positive formulas which possess (at most) the single free variable x_o and are of the form

$$(\exists x_1) \cdots (\exists x_m)(\alpha_1 \wedge \cdots \wedge \alpha_n),$$

where $m,n \in \mathbb{N}$ and $\alpha_1, \ldots, \alpha_n$ are atomic formulas over $\mathfrak{A}$.

Note that for a set of formulas Σ in which only the n variables $x_0, \cdots, x_{n-1}$ occur freely the solvability of Σ in a system $\mathfrak{A}$ is equivalent to the family $\{\pi_n(\mathrm{Sol}_{\mathfrak{A}}(\phi));\ \phi \in \Sigma\}$ having a non-empty intersection, where for $a \in \mathrm{Sol}_{\mathfrak{A}}(\phi)$ $\pi_n(a) := (a(0), \cdots, a(n-1))$, the n-tuple of substitutes for the free variables. Thus, for such systems the solution sets may be viewed as subsets of A^n without ambiguity arising among the relative solvability concepts already introduced. In this vein we formulate

<u>2.5. Definition</u>. Let $\mathfrak{A}$ be an algebraic system and let $n \in \mathbb{N}$. Let $\mathcal{B}_n(\mathfrak{A})$ be the set of subsets of A^n consisting of solution sets of those positive formulas over $\mathfrak{A}$ in which (at most) the variables $x_0, \cdots, x_{n-1}$ occur freely. As $\mathcal{B}_n(\mathfrak{A})$ is closed under finite unions (disjunctions of positive formulas are positive) $\mathcal{B}_n(\mathfrak{A})$ is a base for the closed sets of a topology $\underline{\mathcal{S}_n(\mathfrak{A})}$ on A^n, the <u>n^{th} structure topology on $\mathfrak{A}$</u>.

We will refer to $\mathcal{S}_1(\mathfrak{A})$ simply as the <u>structure topology on $\mathfrak{A}$</u> and write $\underline{\mathcal{S}(\mathfrak{A})}$, or just $\mathcal{S}$, if the intended algebraic system is clear from the context. "Closed" (or "closed in $\mathfrak{A}$") etc. will mean "closed in $\mathcal{S}_n(\mathfrak{A})$", etc., unless reference to another topology is obviously meant. We should remark that in [45] W. Taylor introduced a topology on $\mathfrak{A}$ which, at least in the case that $\mathfrak{A}$ is atomic compact, coincides with $\mathcal{S}(\mathfrak{A})$.

Propositions 3 (i) $\Leftrightarrow$ (ii) and 4 yield immediately:

2.6. Corollary. An algebraic system $\mathfrak{A}$ is atomic compact iff $\mathcal{S}(\mathfrak{A})$ is quasi-compact.

(As a matter of fact atomic compactness is equivalent to the quasi-compactness of any of the nth structure topologies.)

Although the structure topology has, by its very nature, a certain corollation with the algebraic structure, it is in general neither Hausdorff (although it is T_1) nor is it compatible with the algebraic structure in the usual topological algebraic sense. Let us make a few trivial but illustrative observations on the structure topology of a ring R. Although in general the operations + and · are not continuous, the translations $x \mapsto x+a$ and $x \mapsto x \cdot a$ $(a \in R)$ are, as is easily verified - it follows, e.g., that the connected component of zero is an ideal. Also the set of invertible elements of a ring with identity is closed, being the solution set of the formula $(\exists x_1)(x_o \cdot x_1 = 1)$. Thus invertibility is poison for equational compactness : the more invertible elements, the finer the structure topology, as the additive translations are homeomorphisms. In particular, if R is a field, the structure topology is the discrete topology and hence the fact, first observed by J. Mycielski, that an equationally compact field is finite. Looking now at the underlying additive group R^+ of a field R , what can we say? First off Corollary 6 says that $\mathcal{S}(R^+)$ must be quasi-compact, since R^+ is a vector space over R , hence equationally compact as linear spaces are injective. It is not difficult to see that $\mathcal{S}(R^+)$ is in fact the cofinite topology (atomic formulas are now linear equations!) Thus $\mathcal{S}(R^+)$ is neither Hausdorff, nor is addition continuous if $|R| \geq \aleph_0$. This illustrates that, in general, $\mathcal{S}_2(\mathfrak{A})$ is not the product topology $\mathcal{S}_1(\mathfrak{A}) \times \mathcal{S}_1(\mathfrak{A})$, as on R^+ above addition is a continuous map from R^{+2} onto R^+ when R^{+2} is endowed with $\mathcal{S}_2(R^+)$.

Specifically ring theoretic inferences will be taken up again in more detail in Chapter II; let us concentrate for the moment on some universal algebraic consequences.

2.7. Proposition. If $\mathfrak{A} = \langle A; F, R \rangle$ is atomic compact, then so is $\mathfrak{A}^{\mathcal{S}} := \langle A; F, R \cup \mathcal{S}(\mathfrak{A}) \rangle$ (where each member of $\mathcal{S}(\mathfrak{A})$ is given its natural interpretation as a unary relation on $\mathfrak{A}$).

proof: Consider first the algebraic system $\mathfrak{A}^{\mathcal{B}} :=$ $\langle A; F, R \cup \mathcal{B}_1(\mathfrak{A}) \rangle$. It is immediately clear that, in the construction of the basic closed sets of $\mathcal{S}_1(\mathfrak{A}^{\mathcal{B}})$, no new sets are won as were not already present in $\mathcal{B}_1(\mathfrak{A})$. Thus $\mathcal{S}(\mathfrak{A}) = \mathcal{S}(\mathfrak{A}^{\mathcal{B}})$, so $\mathfrak{A}^{\mathcal{B}}$ is atomic compact by Corollary 6. Returning now to $\mathfrak{A}^{\mathcal{S}}$ it is obvious that any atomic formula "$x \in M$" $(M \in \mathcal{S}(\mathfrak{A}))$ occurring in a family of atomic formulas over $\mathfrak{A}^{\mathcal{S}}$ to be tested for the relative solvability property may be replaced by the family $\{x \in M_i;\ i \in I\}$ where M_i are basic closed sets and $M = \bigcap(M_i;\ i \in I)$. Thus atomic compactness of $\mathfrak{A}^{\mathcal{B}}$ implies that of $\mathfrak{A}^{\mathcal{S}}$. ●

What Proposition 7 says is that we can throw in any formulas $x \in M$ where M is closed without violating the relative solvability property. As a matter of fact Proposition 7 may be formulated much more generally : the structure of $\mathfrak{A}$ may be enriched by any and all $\mathcal{S}_n(\mathfrak{A})$-closed n-ary relations on A without losing atomic compactness; for our applications, however, we shall not need this fact.

For an algebra $\mathfrak{A}$ denote by $C_c(\mathfrak{A})$ the set of congruences on $\mathfrak{A}$ which are closed in $\mathcal{S}_2(\mathfrak{A})$.

2.8. Proposition. If $\mathfrak{A}$ is an equationally compact algebra and $\Theta \in C_c(\mathfrak{A})$, then $\mathfrak{A}/\Theta$ is equationally compact.

proof: By hypothesis there is a family $\{\phi_i;\ i \in I\}$ of positive formulas, in which only x_o, x_1 occur freely, such that

$$\Theta = \bigcap(\mathrm{Sol}_{\mathfrak{A}}(\phi_i);\ i \in I) .$$

Now let $\Sigma = \{p_j = q_j;\ j \in J\}$ be a finitely solvable system of polynomial equations over $\mathfrak{A}/\Theta$. To show is that Σ is solvable in $\mathfrak{A}/\Theta$. Let x_{oj}, x_{1j} , $j \in J$, be (pairwise distinct) variables not occurring in any equation in Σ nor in any ϕ_i, $i \in I$. For $j \in J$ construct p_j' from p_j and q_j' from q_j by replacing the occurring constants from A/Θ , in an arbitrary fashion, by their representatives in A . For each $i \in I$ and $j \in J$ let ϕ_{ij} be the formula obtained from ϕ_i by replacing x_o, x_1 by x_{oj}, x_{1j} , resp. It follows that

$$\Omega := \{p_j' = x_{oj};\ j \in J\} \cup \{q_j' = x_{1j};\ j \in J\} \cup \{\phi_{ij};\ i \in I,\ j \in J\}$$

is finitely solvable in $\mathfrak{A}$; namely, solutions are obtained for $p_j' = x_{oj}$, $q_j' = x_{1j}$ by choosing representatives of solutions of the corresponding $p_j = q_j$, and then computing the values of p_j' , q_j' to obtain appropriate substitutes for x_{oj} and x_{1j} . As $\mathfrak{A}$ is positively compact (Proposition 3) Ω is then solvable. Taking the Θ-blocks of any solution of Ω gives a solution of Σ in $\mathfrak{A}/\Theta$. ●

Observe that $C_c(\mathfrak{A})$ forms a complete meet-subsemilattice of $\mathcal{L}(\mathfrak{A})$. In many cases $C_c(\mathfrak{A})$ forms a sublattice, $\mathcal{L}_c(\mathfrak{A})$, of $\mathcal{L}(\mathfrak{A})$, for example when $\mathfrak{A}$ has permutable congruences, as is the case for rings. Note also the simple fact that if $\mathfrak{a}$ is an ideal of a ring R, then $\mathfrak{a}$ is closed in R iff the congruence relation induced by $\mathfrak{a}$ is $\mathcal{J}_2$(R)-closed. Thus the notion "closed ideal" is unambiguous. The same is, of course, true for "closed submodule" and in particular "closed one-sided ideal" : the interpretation as subset or as congruence relation is irrelevant. The following result is a further topology-like feature.

2.9. Proposition. Let $\mathfrak{B} = \langle B; F', R' \rangle$ be a subreduct of the algebraic system $\mathfrak{A} = \langle A; F, R \rangle$. If $\mathfrak{A}$ is atomic compact and B is closed in $\mathfrak{A}$, then $\mathfrak{B}$ is atomic compact.

proof: Let $\mathfrak{A}' = \langle A; F', R' \rangle$ be the reduct of $\mathfrak{A}$ such that $\mathfrak{B} \leq \mathfrak{A}'$. Let Σ be a system of α-ary atomic formulas over $\mathfrak{B}$, finitely solvable in $\mathfrak{B}$. Then $\Sigma \subseteq At_\alpha^B(\mathfrak{A}')$. By the very definition of "reduct" we may assume that $\Sigma \subseteq At_\alpha^B(\mathfrak{A})$, as only the question of solvability is at stake. It follows that

$$\Sigma' = \Sigma \cup \{x_\gamma \in B;\ 0 \leq \gamma < \alpha\}$$

is finitely solvable in $\mathfrak{A}$. As B is closed in $\mathfrak{A}$ and $\mathfrak{A}$ is atomic compact, Σ' is solvable in $\mathfrak{A}$ by Proposition 7, and any solution is, of course, a solution of Σ in $\mathfrak{B}$. •

On most occasions when applying the above a particular case will suffice:

2.10. Corollary. If $\mathfrak{A}$ is atomic compact and $\mathfrak{B} \subseteq \mathfrak{A}$ such that B is closed in $\mathfrak{A}$, then $\mathfrak{B}$ is atomic compact.

In topological algebra the question as to the embeddability of an algebra into a compact one is, next to the question of compactness itself, of vital interest. And with a view towards this question the corresponding theory of "(equational) compactifications" has its obvious place, and is therefore of inherent interest. That part of the theory which will be of later use to us we sketch below. Adopting the terminology of [55] (notation varies in the literature) we make the following

2.11. Definition. Let $\mathfrak{A}$ be a subsystem of the algebraic system $\mathfrak{B}$.

(1) $\mathfrak{B}$ is a compactification of $\mathfrak{A}$ if $\mathfrak{B}$ is atomic compact.

(2) $\mathfrak{B}$ is a closure of $\mathfrak{A}$ if $\mathfrak{B}$ is $At^A(\mathfrak{B})$-compact.

(3) $\mathfrak{B}$ is a quasi-compactification of $\mathfrak{A}$ if any set of formulas from $At^A(\mathfrak{A})$ which is finitely solvable in $\mathfrak{A}$ is solvable in $\mathfrak{B}$.

Call an algebraic system $\mathfrak{A}$ (quasi-) compactifiable if a (quasi-) compactification of $\mathfrak{A}$ exists. Denote by $\mathbf{c}(\mathfrak{A})$ the class of quasi-compactifications of $\mathfrak{A}$.

(For the development of a theory of equational compactifications and "equationally compact hulls" of a more category theoretical bent - which will not be touched upon in this treatise - we refer to [3] and [4].) Of the above concepts, listed in order of decreasing strength, (1) and (3) were introduced and first studied by B. Weglorz in [54]; there he proved that if an algebra $\mathfrak{A}$ possesses a (quasi-) compactification at all, it possesses one in the equational class generated by $\mathfrak{A}$. Later G. H. Wenzel substantially improved this result, and it is this sharpened form which will prove invaluable in the next chapter.

2.12. Proposition (G.H. Wenzel [57]). If the algebraic system $\mathfrak{A}$ is (quasi-) compactifiable, then there is a (quasi-) compactification $\mathfrak{B}$ of $\mathfrak{A}$ such that every positive sentence over $\mathfrak{A}$ which holds in $\mathfrak{A}$ holds also in $\mathfrak{B}$.

In [54] B. Weglorz posed a question which remains unanswered to this day - the *Weglorz Problem* : does every quasi-compactifiable algebra possess a compactification? The attractiveness of a positive answer to this question is clear : investigating the *quasi*-compactifiability of an algebra requires an analysis of all systems of equations over $\mathfrak{A}$ which are finitely solvable in $\mathfrak{A}$, a job demanding only sufficient "knowledge" of $\mathfrak{A}$; whereas determining the existence of a compactification or closure demands a similar analysis, but one for *each extension* of $\mathfrak{A}$ as well, a priori a much more involved task. Nominal progress has been made, though, towards a positive answer. W. Taylor [46] proved that an algebra with a closure is compactifiable, which of course eases the task, at least on the surface, to some degree. In a different vein G. H. Wenzel (unpublished) proved that $\mathfrak{A}$ is compactifiable as soon as every member of $\mathbf{HSP}(\mathfrak{A})$ is quasi-compactifiable. Although subalgebras and products of quasi-compactifiables are quasi-compactifiable, homomorphic images do not in general inherit this property, so, unfortunately, this result does not yield directly a solution of the problem either.

In Chapter II we shall deal in particular with this question in rings; in some classes of rings, it turns out, the quasi-compactifiable ones are already equationally compact. We conclude the section and the chapter with a useful criterium in the presence of which quasi-compactifications totally lack. Although it is tailored to the ring theoretic needs of Chapter II it is of a universal algebraic nature, and so we present it here.

2.13. Proposition. Let $\mathfrak{A}$ be an algebraic system. Suppose there are distinct elements 0 and c in A, an infinite subset S of A, a subset T of A, a binary polynomial d over $\mathfrak{A}$, and an $(n+1)$-ary polynomial p over $\mathfrak{A}$ satisfying

(1) $d(a,a) = 0$, for $a \in A$ and $d(s,t) \in T$ for all $s \neq t$ in S,

(2) for each $t \in T$ there are $b_1,\dots,b_n \in A$ such that $p((t,b_1,\dots,b_n) = c$ but $p(0,b_1,\dots,b_n) = 0$ for all $b_1,\dots,b_n \in A$.

Then $\mathbf{c}(\mathfrak{A}) = \emptyset$.

proof: Suppose not. Then by Proposition 12 there is $\mathfrak{B} \in \mathbf{c}(\mathfrak{A})$ satisfying every positive sentence over $\mathfrak{A}$ holding in $\mathfrak{A}$, and in particular every polynomial identity over $\mathfrak{A}$ which is satisfied by $\mathfrak{A}$. Let I be a set such that $|I| > |B|$. Consider the system of equations

$$\Sigma = \{x_{ij} = d(z_i,z_j);\ i,j \in I,\ i \neq j\} \cup \{p(x_{ij},y_{ij}^{(1)},\dots,y_{ij}^{(n)}) = c;\ i,j \in I,\ i \neq j\},$$

where all the x's, y's and z's are pairwise distinct variables. Then in any finite subset of Σ the occurring z_i's (finitely many only!) can be substituted by pairwise distinct elements of S. Plugging these values into equations of the type

$x_{ij} = d(z_i,z_j)$ we get values for the x_{ij}'s which by (1) lie in T. Putting these x_{ij}'s into equations of the remaining type (right), condition (2) guarantees the existence of substitutes for the $y_{ij}^{(k)}$'s yielding a solution of these. Thus Σ is finitely solvable in $\mathfrak{A}$, and hence solvable in $\mathfrak{B}$. Now the polynomial identities $d(x,x) = 0$ and $p(0,x_1,\cdots,x_n) = 0$ hold in $\mathfrak{A}$ by (1) and (2), and hence in $\mathfrak{B}$. But in any solution of Σ the substitutes for at least two distinct z_i, z_j must agree by the cardinality condition on I; hence some x_{ij} must assume the value 0, making the simultaneous solvability of $p((x_{ij},y_{ij}^{(1)},\cdots,y_{ij}^{(n)}) = c$ impossible - a contradiction. ●

CHAPTER II.

MINIMUM CONDITIONS AND COMPACTIFICATIONS

§3 Structural conditions for equational compactness and quasi-compactifiability; ring extensions

In this section we derive a number of results of the type indicated in the title but not depending on minimum conditions on the ring per se. They will not only have direct applications in the discussion of the next paragraph, focussing on minimum conditions, but will also find important uses in Chapter III dealing with rings with the maximum condition.

All rings are associative. We use the customary abbreviations A.C.C. (D.C.C.) for the ascending (descending) chain condition on left ideals ("Ideal" unmodified always means two-sided ideal). We call a ring with D.C.C. also artinian; a noetherian ring will denote a ring possessing an identity and satisfying A.C.C. The characteristic of a ring R, $\chi(R)$, is defined to be the torsion bound of R^+ if R^+ is a bounded torsion group, and zero otherwise. $J(R)$ will denote the Jacobson radical of R and a ring is semisimple if $J(R) = (0)$. Recall that $J(R)$ is the largest left-quasi-regular left ideal of R, that is, $J(R)$ consists precisely of those elements of R which generate principal left-quasi-regular left ideals, i.e., precisely those elements in R which are solutions, for each $s \in R$ and $z \in \mathbb{Z}$, of the positive formulas

$$\phi_{s,z} \equiv (\exists x_1)((s\cdot x_o + z\cdot x_o) + x_1 + x_1(s\cdot x_o + z\cdot x_o) = 0)$$

over R . This proves

3.1. Proposition. $J(R)$ is closed in R .

As elementary as the above observation may be, it does indicate a certain naturalness in the structure topology vis à vis topologies compatible with R in the usual sense; Indeed, in [28] I. Kaplansky exhibited a topological ring in which $J(R)$ is not closed; and although in a compact topological ring $J(R)$ is closed [28, p.163] the proof of this fact is anything but elementary. We record for later reference another simple, but important, observation:

3.2. Proposition. Let R be a ring which either has an identity or is such that R^+ is a torsion group. Then every finitely generated left ideal is closed.

proof: Let $\mathfrak{a}$ be a left ideal of R generated, say, by $a_1,\dots,a_n$. If R is unital, set $k = 1$, and if R^+ is torsion, let k be a natural number at least as large as the order of each a_i , $i = 1,\dots,n$. Then $\mathfrak{a}$ is the solution set (in both cases) of the single formula

$$(\exists x_1)\cdots(\exists x_n)\bigvee_{\substack{0\le m_i<k\\1\le i\le n}}(x_0 = x_1a_1+\cdots+x_na_n+m_1a_1+\cdots+m_na_n) .$$

Hence $\mathfrak{a}$ is closed in R . ●

By way of contrast let us give an example of an ideal of a ring which is not closed : Let, for every $n\in\mathbb{N}$, R_n be some finite unital ring, and set $R = \prod(R_n;\ n\in\mathbb{N})$. Then

R is equationally compact, being compact. Now $\bigoplus(R_n;\ n \in \mathbb{N})$ is not equationally compact [20, Prop. 4.12]. But then $\bigoplus(R_n;\ n \in \mathbb{N})$, viewed in the canonical fashion as an ideal of R, is not closed, as otherwise Corollary 2.10 is contradicted. This example also illustrates that, although for any ring S $\mathcal{L}_c(S)$ is a complete lattice (and a sublattice of $\mathcal{L}(S)$), $\mathcal{L}_c(S)$ is not in general a complete sublattice of $\mathcal{L}(S)$; for $\bigoplus(R_n;\ n \in \mathbb{N})$ is the supremum in $\mathcal{L}(R)$ of the family of (closed!) ideals $\{R_n;\ n \in \mathbb{N}\}$ but of course not the supremum in $\mathcal{L}_c(R)$. (In this case R is the supremum in $\mathcal{L}_c(R)$ by [20, Prop. 4.12] and 2.10 again.)

A ring is a structural enrichment of an abelian group; hence a ring R is equationally compact only if its reduct R^+ is also. Equationally compact abelian groups, however, have a nice description. This is

<u>3.3. Definition (I. Kaplansky [29])</u>. An abelian group G is <u>algebraically</u> <u>compact</u> if

$$G \cong D \oplus \prod(G_p;\ p \in \mathbb{P}),$$

where D is divisible, $\mathbb{P}$ denotes the set of prime natural numbers, and for each $p \in \mathbb{P}$ G_p is a module over the p-adic integers Z_p^*, the p-adic topology on G_p is Hausdorff, and G_p is complete in it.

J. Łoś proved in [34] that equational compactness in abelian groups is characterized by algebraic compactness, and we shall exploit this property in equationally compact rings. To this end, the following internal description of the factors of an algebraically compact group will prove useful.

<u>3.4. Proposition (I. Kaplansky [29])</u>. If $G = D \oplus \prod(G_p;\ p \in \mathbb{P})$ is an algebraically compact group, then - viewing the factors as subgroups in the canonical fashion - D is the largest divisible subgroup of G, and for each prime number p, G_p is the largest subgroup of G divisible by all powers of all primes excepting p.

The next result exploits the algebraic compactness of the abelian group underlying an equationally compact ring.

<u>3.5. Proposition</u>. Let R be an equationally compact ring such that R^+ is a reduced group. Then

$$R \cong \prod(R_p;\ p \in \mathbb{P}),$$

where each R_p is an algebra over the p-adic integers $\mathbb{Z}_p^*$, complete in the p-adic topology, which is Hausdorff.

<u>proof</u>: The hypothesis on R together with 3.3 imply that

$$R^+ \cong \prod(R_p;\ p \in \mathbb{P})$$

where each R_p is a $\mathbb{Z}_p^*$-module with the desired properties. Identify the carrier set R with $\prod(R_p;\ p \in \mathbb{P})$. If we can show that multiplication in R is componentwise multiplication in $\prod(R_p;\ p \in P)$, then the given (group) decomposition of R^+ will also be ring direct. So let

$$\bar{a}_p := (a_p)_{p \in \mathbb{P}} \qquad \text{and} \qquad \bar{b}_p := (b_p)_{p \in \mathbb{P}}$$

be arbitrary elements of $\prod(R_p;\ p \in \mathbb{P})$. Fix $q \in \mathbb{P}$ and write $\bar{a}_p = a_q + (a_p)_{p \neq q}$ and $\bar{b}_p = b_q + (b_p)_{p \neq q}$ (where $a_q :=$ $(c_p)_{p \in \mathbb{P}}$ with $c_p = 0$ for $p \neq q$, $c_q = a_q$, and $(a_p)_{p \neq q} :=$ $(d_p)_{p \in \mathbb{P}}$ with $d_p = a_p$ for $p \neq q$, $d_q = 0$; similarly with "b"); then compute :

$$\bar{a}_p \cdot \bar{b}_p = a_q \cdot b_q + a_q \cdot (b_p)_{p \neq q} + (a_p)_{p \neq q} \cdot b_q + (a_p)_{p \neq q} \cdot (b_p)_{p \neq q}$$

By Proposition 4 R_p is the largest subgroup H of R^+ with the following property: every element of H is divisible (in H) by all powers of every prime not equal to p . From this and the above decomposition it is clear that R_p is the set of all elements which are divisible <u>in R^+</u> by all powers of every prime not equal to p . Moreover, since R_p has no non-zero elements divisible by every power of p , no non-zero element of R can be divisible in R^+ by every power of every prime. Together this implies that $a_q \cdot b_q \in R_q$, $a_q \cdot (b_p)_{p \neq q}$ = $(a_p)_{p \neq q} \cdot b_q$ = 0 , and that the q-th component of $(a_p)_{p \neq q} \cdot (b_p)_{p \neq q}$ is zero. Thus we have shown that, if $\bar{a}_p \cdot \bar{b}_p$ = $\bar{c}_p$, then for each prime q , $c_q = a_q \cdot b_q$, i.e., multiplication is componentwise.

It remains only to show that the subrings R_p are $\mathbb{Z}_p^*$-algebras, i.e., that, in addition, $(z \cdot r) \cdot s = z \cdot (r \cdot s) = r \cdot (z \cdot s)$ holds for arbitrary $z \in \mathbb{Z}_p^*$ and $r, s \in R_p$. But $\mathbb{Z}_p^*$ is just the completion of $\mathbb{Z}$ with respect to the p-adic topology on $\mathbb{Z}$, and since R_p is a $\mathbb{Z}$-algebra with Hausdorff p-adic topology the above identities are obtained by lifting in the standard fashion the same identities which hold down on the $\mathbb{Z}$-algebra R_p . ●

For a ring R and a natural number $n > 1$ let R_n denote the ring of $n \times n$ matrices over R .

<u>3.6. Proposition</u>. A unital ring R is quasi-compactifiable (equationally compact) if and only if R_n is quasi-compactifiable (equationally compact).

proof: Suppose R is quasi-compactifiable; by Proposition 2.12 there is a ring $S \in \mathbf{c}(R)$. We claim that $S_n \in \mathbf{c}(R_n)$. Replacing each variable x occurring in an equation ϕ over R_n by the variable matrix $(x_{ij})_{1 \leq i,j \leq n}$ and multiplying out, we obtain a matrix $(\phi_{ij})_{1 \leq i,j \leq n}$ of n^2 equations over R. Obviously a system of equations over R_n, finitely solvable in R_n, reduces in the above fashion to a finitely solvable system over R, which is then solvable in S by hypothesis, and the solution matrices clearly are solutions of the original system in S_n. Supposing that R is equationally compact, the proof that R_n is equationally compact is totally analogous to the above.

Conversely, assume R_n is quasi-compactifiable; as R is canonically embedable in R_n, any $S \in \mathbf{c}(R_n)$ is trivially a quasi-compactification of R. Finally, let R_n be equationally compact, and let for any $r \in R$ $\hat{r} \in R_n$ be the matrix with r in the upper left hand corner and zero's elsewhere. If Σ is a system of equations over R, finitely solvable in R, obtain the the system $\hat{\Sigma}$ of equations over R_n by replacing all constants r by $\hat{r}$ and all occurrences of all variables x by $\hat{1} \cdot x \cdot \hat{1}$. Clearly $\hat{\Sigma}$ is finitely solvable in R_n, hence solvable in R_n, and the upper left hand entries of solutions of $\hat{\Sigma}$ in R_n give solutions of Σ in R. •

For any ring R denote by $D(R)$ the largest divisible subgroup of R^+, and let $\mathrm{Ann}(R)$ denote the annihilator of R: $\mathrm{Ann}(R) = \{r \in R\ ;\ r \cdot x = x \cdot r = 0 \text{ for all } x \in R\}$.

3.7. Proposition. If R is a quasi-compactifiable ring, then $D(R) \subseteq \mathrm{Ann}(R)$; in particular, $D(R)$ is an ideal. If, moreover,

R is equationally compact, then D(R) is closed.

<u>proof</u>: Assume that $s \cdot d \neq 0$ for some $s \in R$ and $d \in D(R)$. Apply Proposition 2.13 with $c := s \cdot d$, $T = S := \mathbb{Z} \cdot s \setminus \{0\}$, $d(x,y) := x-y$, and $p(x_o, x_1) := x_o \cdot x_1$. Since for any $z \in \mathbb{Z} \setminus \{0\}$ there is by divisibility $d' \in D(R)$ satisfying $z \cdot d = d$ and hence $(z \cdot s) \cdot d' = s \cdot (z \cdot d') = s \cdot d$ ($\neq 0$!), it follows, first, that S is infinite and, second, that conditions (1) and (2) of 2.13 are satisfied. Thus $c(R) = \emptyset$. This contradiction shows that $R \cdot D(R) = \{0\}$; similarly $D(R) \cdot R = \{0\}$. If R is equationally compact then R^+ is algebraically compact and it is then evident from Definition 3 and Proposition 4 that D(R) is the <u>set</u> of all elements in R which are divisible <u>in R</u> by all integers. It follows that

$$D(R) = \bigcap \mathrm{Sol}_R((\exists x_1)(x_o = z \cdot x_1); z \in \mathbb{Z})$$

and so D(R) is closed in R . ●

Together with Proposition 2.8 the above yields

<u>3.8. Corollary</u>. If R is an equationally compact ring then D(R) is an ideal and R/D(R) is equationally compact.

Another immediate application gives

<u>3.9. Corollary</u>. An equationally compact semisimple ring R has the decomposition given in Proposition 5.

<u>proof</u>: $D(R)^2 = (0)$ and so D(R) is a nilpotent ideal and hence contained in J(R) , which is (0) . Thus R^+ is reduced and Proposition 5 applies. ●

With a glance towards I. Kaplansky's structure theorem for compact semisimple rings [28, Theorem 16], or towards D. Zelinsky's structure theorem for linearly compact semisimple rings [61, Prop. 11], it is natural to ask, in light of the decomposition given above, whether an equationally compact semisimple ring isn't already compact, or at least linearly compact (in any one of the senses flourishing in the literature). That this is not so will be illustrated by an example given in §7.

Suppose that A and R are rings such that A has an identity and R is an A-algebra which is unital as a (left) A-module. Denote by $A*R$ the ring obtained by the standard adjunction of an identity construction. That is, the carrier set is $A \times R$, addition is componentwise, and multiplication is given by $(a,r)\cdot(b,s) = (a\cdot b,\ a\cdot s+b\cdot r+r\cdot s)$. We identify A and R with their images under the canonical (ring) embeddings into $A*R$. Note that $A*R$ has an identity, R is an ideal of $A*R$, and $A*R$ satisfies A.C.C., or D.C.C., as soon as both A and R do the same.

3.10. Proposition. Let R be an A-algebra (in the above sense). If A is finite and R is equationally compact (quasi-compactifiable) as an A-algebra then the ring $A*R$ is equationally compact (quasi-compactifiable).

proof: We prove the statement in parentheses. Assume that S is an A-algebra quasi-compactifying R (2.12 again). We claim that $A*S$ is a quasi-compactification of $A*R$. Now if ϕ is an α-ary polynomial equation over $A*R$, then by replacing every variable x_γ occurring in ϕ by the variable pair (z_γ, y_γ) and calculating out according to the rule defining the operations, ϕ is then transformed into a pair (ϕ_1, ϕ_2) of equations, where ϕ_1 is an α-ary ring equation in the z_γ's over A and ϕ_2 is a "generalized" α-ary A-algebra equation with constants in R, in the variables x_γ and <u>scalar</u> variables z_γ, $0 \leq \gamma < \alpha$. Furthermore, if the substitution of $(a,r) \in A \times R$ for the variable x_γ in ϕ is interpreted in (ϕ_1, ϕ_2) by the substitution of a for z_γ and r for y_γ, then all solvability properties are left invariant by this transformation process.

Let Σ be a system of α-ary polynomial equations over $A*R$, finitely solvable in $A*R$. We may assume that the members of Σ have been transformed via the process described above. The idea now is to reduce Σ to a system of A-algebra equations over R, by replacing one by one the z_γ's by elements of A without violating the finite solvability along the way. The formal argument runs as follows: For an ordinal $\gamma < \alpha$ and a γ-tuple $\underline{a} \in A^\gamma$ let $\Sigma_{\gamma,\underline{a}}$ be the system of equations obtained from Σ by replacing each occurrence of z_β, $0 \leq \beta < \gamma$, in each member of Σ by $\underline{a}(\beta)$. Let $\mathcal{F}$ be the family of all those systems $\Sigma_{\gamma,\underline{a}}$ which are finitely solvable in $A*R$. $\mathcal{F}$ is not empty, as $\Sigma = \Sigma_{0,\emptyset} \in \mathcal{F}$. Partially order $\mathcal{F}$ by

$$\Sigma_{\gamma,\underline{a}} \leq \Sigma_{\gamma',\underline{a}'} \quad : \text{iff} \quad \gamma \leq \gamma' \quad \text{and} \quad \underline{a}(\beta) = \underline{a}'(\beta) \text{ for all } \beta < \gamma.$$

It follows immediately that $\mathcal{F}$ is inductive (finite solvability demands looking at only a finite number of variables at a time) and hence has a maximal member, say $\Sigma_{\delta,\underline{a}}$, by Zorn's Lemma. We show that $\delta = \alpha$. Suppose not. Defining, for $c \in A$, the

the $\delta+1$-tuple $\underline{a}_c$ by $\underline{a}_c(\delta) = c$ and $\underline{a}_c(\gamma) = \underline{a}(\gamma)$ for $\gamma < \delta$, it follows by the maximality of of $\Sigma_{\delta,\underline{a}}$ that $\Sigma_{\delta+1,\underline{a}_c}$ is not finitely solvable for any $c \in A$. But since A is finite, this means there is a finite subset of $\Sigma_{\delta,\underline{a}}$ such that for no substitution of z_δ by an element of A is the resulting system solvable, contradicting the finite solvability of $\Sigma_{\delta,\underline{a}}$. Thus $\delta = \alpha$ as claimed. But now $\Sigma_{\delta,\underline{a}}$ consists of pairs of equations (ϕ_1,ϕ_2) where ϕ_1 is a constant identity $a = a$ $(a \in A)$ and ϕ_2 is an A-algebra equation over R; the system of ϕ_2's is finitely solvable in R, hence solvable in S, and if $\underline{s} \in S^\alpha$ is a solution, then clearly $\underline{u} \in (A\times S)^\alpha$ defined by $\underline{u}(\gamma) = (\underline{a}(\gamma), \underline{s}(\gamma))$, $0 \le \gamma < \alpha$, is a solution of Σ in $A*S$. Thus $A*S \in \mathbf{c}(A*R)$. The proof of the "equationally compact" statement is obtained by putting $S = R$ in the above argument. ●

If R is a ring with positive characteristic n, then R is a Z_n-algebra, where Z_n denotes the ring of integers modulo n. Moreover, if R is equationally compact, then R is an equationally compact Z_n-algebra, as Z_n-algebra polynomials are already ring polynomials. Summarizing, we have the important

<u>3.11. Corollary</u>. If R is an equationally compact (quasi-compactifiable) ring of characteristic $n > 0$, then Z_n*R is equationally compact (quasi-compactifiable).

We conclude with an interesting application of the results of this section characterizing the equationally compact rings which are subrings of equationally compact rings with identity.

3.12. Theorem. Let R be an equationally compact ring. Then R is a subring of an equationally compact ring with identity if and only if R^+ is a reduced group.

proof: Note first the elementary fact that a direct product of rings is equationally compact if and only if each factor ring is. Now if R^+ is reduced, then by Proposition 5 $R \cong \prod(R_p;\ p \in \mathbb{P})$ such that for each p, $\bigcap(p^n R_p;\ n \in \mathbb{N}) = (0)$. Now R_p is an equationally compact ring (being a factor of one) and for each n, $p^n R_p$ is an ideal in R_p, which, moreover, is closed, being the solution set of $(\exists x_1)(x_o = p^n x_1)$; hence $R_p/p^n R_p$ is an equationally compact ring by Proposition 2.8. Since it also has positive characteristic it is a subring of an equtionally compact ring with identity, say S_{p^n}, by Corollary 11. By way of the canonical projections of R_p onto $R_p/p^n R_p$ we therefore obtain embeddings

$$R_p \rightarrowtail \prod(R_p/p^n R_p;\ n \in \mathbb{N}) \rightarrowtail \prod(S_{p^n};\ n \in \mathbb{N}) =: S_p .$$

Thus R_p is embeddable in the equationally compact ring with identity S_p. Taking products again we get

$$R \cong \prod(R_p;\ p \in \mathbb{P}) \rightarrowtail \prod(S_p;\ p \in \mathbb{P}) =: S ,$$

and S is an equationally compact ring with identity.

Conversely, suppose $R \subseteq S$, where S is equationally compact and has an identity. Any divisible subgroup of R^+ must annihilate all of S by Proposition 7, i.e., must be (0) as S has an identity. Thus R^+ is reduced. ●

§4 Compactifications of artinian rings and simple rings with minimal ideals

This section is devoted to the problem of characterizing within various classes of rings those members having quasi-compactifications. Satisfactory answers can be given in some cases, most notably in the class of artinian rings. In the following $Z(R)$ denotes the center of the ring R. The heart of a subdirectly irreducible ring R (its unique smallest non-zero ideal) denote by $Ht(R)$. In a simple ring R, $R^2 \neq (0)$ is always to hold. We begin with some necessary conditions for a subdirectly irreducible ring to be quasi-compactifiable.

4.1. Proposition. Let R be a quasi-compactifiable subdirectly irreducible ring. Then :

(i) The ring $Z(R)/Ann(R)$ is finite.

(ii) If $\chi(R) = 0$ $D(R)$ is isomorphic to the Prüfer group $\mathbf{Z}(p^\infty)$ for some prime p ($\mathbf{Z}(p^\infty)$ is the multiplicative group of all complex p^n-th roots of unity, $n \in \mathbb{N}$).

proof: (i). We assume that $Z(R)/Ann(R)$ is infinite and apply Proposition 2.13 : To this end, fix c to be an arbitrary non-zero element in $Ht(R)$, let $d(x,y) = x - y$, $p(x_0,x_1) = x_0 \cdot x_1$, let S be a complete system of representatives of $Z(R)/Ann(R)$, and let T be the set $Z(R) \setminus Ann(R)$. Clearly condition (1) of 2.13 holds; for condition (2) note that, for $t \in T$, $t \cdot R$ is a non-zero (two-sided!) ideal of R, hence it contains $Ht(R)$, and so c has a representation as $t \cdot b$ for some $b \in R$. Thus 2.13 yields the desired contradiction.

(ii). For any subdirectly irreducible ring R , R^+ is reduced iff $\chi(R) > 0$, because $\{n \cdot R;\ n \in \mathbb{N}\}$ is a family of ideals whose meet is divisible. Thus $D(R) \neq (0)$ in the case at hand. As is well-known (see [29]) D(R) is the direct sum of Prüfer groups and copies of the rationals $\mathbb{Q}$; since D(R) is contained in Ann(R) by Proposition 3.7 each of the summands occurring is an ideal of R ; thus by subdirect irreducibility this representation of D(R) can have only one summand, which cannot be $\mathbb{Q}$ because $\mathbb{Q}$ has a family of non-zero subgroups (= ideals in R) meeting in (0) , which fact would again contradict subdirect irreducibility. •

Observe that the situation described in (ii) may occur : the group $\mathbb{Z}(p^\infty)$ is itself subdirectly irreducible and divisible (hence equationally compact); thus the zero-ring $\mathbb{Z}(p^\infty)_0$ is an equationally compact subdirectly irreducible ring. From (i) follows immediately :

<u>4.2. Corollary</u>. A quasi-compactifiable field is finite. And thus the center of a quasi-compactifiable simple ring is finite (being either (0) or a field).

Our immediate goal is a sharpening of Theorem 3.12 for the case that R is subdirectly irreducible. To this end we improve (i) of Proposition 1 for the case that R^+ is reduced.

<u>4.3. Proposition</u>. Let R be equationally compact and subdirectly irreducible. If $\chi(R) > 0$ then Z(R) is finite.

<u>proof</u>: Let $\chi(R) = m$ (of course m is actually a prime power

as R is subdirectly irreducible), and suppose that $Z(R)$ is infinite. Consider the following set of disjunctions of equations :

$$\Sigma := \{ \bigvee (k(x_i - x_j) + (x_i - x_j)y_{ij} = c \ ; \ 1 \leq k \leq m); \ i,j \in I, \ i \neq j \} \ ,$$

where $c \in Ht(R)$, $c \neq 0$, and I is a set such that $|I| > |R|$. Σ is finitely solvable; indeed, the right ideal generated by any non-zero central element r has the form $\mathbb{Z} \cdot r + r \cdot R$; as it is also two-sided it contains $Ht(R)$ and hence c - thus any finite subset of Σ can be solved by replacing the occurring x_i's be pairwise distinct members of $Z(R)$. Since R is positively compact, Σ is solvable. However in any solution of Σ the cardinality of I forces two distinct x_i's to take on the same value, implying that $c = 0$, a contradiction. ●

Remark. The above argument smacks of Proposition 2.13; there a polynomial p (yielding an equation satisfying certain properties) is required, whereas above we have a disjunction of equations sporting the same properties. Although not stated explicitly, it is implicit in the proof of Theorem 2.2.1 (6)$\Rightarrow$(1) of G.H. Wenzel [57] that $\mathfrak{B}$ is a quasi-compactification of the algebraic system $\mathfrak{A}$ only if every set of positive formulas over $\mathfrak{A}$ not involving the quantifier $(\forall x)$ is solvable in $\mathfrak{B}$ provided it is finitely solvable in $\mathfrak{A}$. With the help of 2.12 this fact does allow a more general formulation of Proposition 2.13 so that Proposition 3 formulated with the weaker hypothesis "quasi-compactifiable" falls out as an application. As Proposition 3 is an isolated case where a stronger form of 2.13 is required, and whereas a sharpening of Proposition 3 would not help anyway in sharpening Theorem 4 below, we have refrained from expounding on a generalization along these lines. With the help of Proposition 3 above we achieve a specialization of Theorem 3.12.

4.4. Theorem. For an equationally compact subdirectly irreducible ring R the following are equivalent:

(i) R is a subring of an equationally compact subdirectly irreducible ring T with identity.

(ii) R^+ is reduced.

If R is commutative then the T in (i) can also be chosen commutative.

proof: (i)⇒(ii) follows directly from Theorem 3.12.
(ii)⇒(i): Let $\chi(R) = p^m$ ($p \in \mathbb{P}$, $m \in \mathbb{N}$). We have then R canonically embedded into the ring $S := \mathbb{Z}_{p^m} * R$, which is equationally compact by Corollary 3.11. Observe that every ideal of R is an ideal of S and that, via the canonical embeddings, the element $(n,r) \in S$ is identified with $n+r$. Suppose $\mathfrak{a}$ is an ideal in S satisfying $\mathfrak{a} \cap R = (0)$, and let $n+r \in \mathfrak{a}$ with $n \in Z_{p^m}$ and $r \in R$. For any $s \in R$ we have then

$$s(n+r) = ns + sr \in \mathfrak{a} \cap R = (0) ,$$

i.e., $s(n+r) = 0$. Similarly, $(n+r)s = 0$. It follows that $sr = -ns = rs$ and therefore $r \in Z(R)$ as s was arbitrary. Hence $\mathfrak{a} \subseteq Z_{p^m} + Z(R)$. But $Z(R)$ is finite by Proposition 3 and we conclude that there are only a finite number of ideals $\mathfrak{a}$ satisfying $\mathfrak{a} \cap R = (0)$, and each of these is finite. Choose an ideal $\mathfrak{a}$ of S maximal with respect to the property $R \cap \mathfrak{a} = (0)$. Setting $T = S/\mathfrak{a}$ it follows that R is embedded into T via the projection, and we claim T does the job.

First off $\mathfrak{a}$ is finite and hence closed, so T is equationally compact by Proposition 2.8 and Corollary 3.11. It remains to show that T is subdirectly irreducible (of course T has an identity because S has one). Identifying R with its image under the projection into T, $Ht(R)$ is an ideal of T because $Ht(R)$ is an ideal of S and T is a homomorphic image of S; for the same reason R is an ideal of T, and as every non-zero ideal of T meets R non-trivially by the maximality of $\mathfrak{a}$, every non-zero ideal of T must contain $Ht(R)$. We conclude that T is subdirectly irreducible with $Ht(T) = Ht(R)$. Of course, if R is commutative, then T as constructed above is so too. ●

Before losing sight of the above proof let us record

<u>4.5. Corollary</u>. An equationally compact simple ring R without an identity is a subring of an equationally compact ring T with identity in which R is the only non-trivial ideal.

<u>proof</u>: Observe first that $\chi(R) = p$ for some $p \in \mathbb{P}$ as otherwise $pR = R$ for each $p \in \mathbb{P}$, as pR is an ideal, and so R^+ is divisible and thus $R^2 = (0)$ by Proposition 3.7, a contradiction. For the rest, observe that, following the proof of the above for this case, $Ht(T) = Ht(R) = R$; but there are no ideals lying properly between R and $S = Z_p*R$, hence none lying properly between $Ht(T)$ and T. Of course $R \neq T$ as T has an identity and R does not. ●

One precipitate of the foregoing investigations is that in the variety of commutative unital rings all equationally compact subdirectly irreducible members are finite. However this fact aids in structural considerations of equationally compact members of this variety only when one has the guarantee that subdirectly irreducible quotients of equationally compact members are again equationally compact, as is the case for noetherian rings. This situation is covered by the discussion in Chapter III.

In some cases the presence of a socle (i.e., of a minimal one-sided ideal) is definitive for the question of quasi-compactifiability, as is illustrated by the following observations.

4.6. Proposition. A quasi-compactifiable simple ring R with socle is finite.

proof: By simplicity the right annihilator of R is (0). It follows that for any non-zero left ideal $\mathfrak{a}$ of R the left annihilator $\mathrm{Ann}_1(\mathfrak{a})$ cannot be R, hence (being an ideal) must be (0). Let $\mathfrak{a}$ be a minimal left ideal of R; then fixing $c \in \mathfrak{a}$, $c \neq 0$, it follows that $\mathfrak{a} = R\cdot c$ since $R\cdot c \neq (0)$ and $R\cdot c$ is a left ideal contained in $\mathfrak{a}$. As $\mathrm{Ann}_1(R\cdot c) \neq (0)$ we have that for any $x_0 \neq 0$ there is $x_1 \in R$ such that $x_0\cdot x_1\cdot c \neq 0$. But then as above $R\cdot x_0\cdot x_1\cdot c = \mathfrak{a}$ as $x_0\cdot x_1\cdot c$ is a non-zero element in $\mathfrak{a}$. We conclude that for any $x_0 \neq 0$ there are x_1 and x_2 in R satisfying

$$x_2\cdot x_0\cdot x_1\cdot c = \bar{c} .$$

Thus if we assume that R is infinite the conditions of Proposition 2.13 are satisfied with $c \in \mathfrak{a}\setminus\{0\}$, $d(x,y) = x - y$, $T = S = R\setminus\{0\}$ and

$$p(x_0,x_1,x_2) = x_2\cdot x_0\cdot x_1\cdot c ,$$

a contradiction. ●

In a similar vein we have

4.7. Proposition. A quasi-compactifiable integral domain R with socle is finite.

proof: Letting $\mathfrak{a}$ be a minimal left ideal and fixing $c \in \mathfrak{a} \setminus \{0\}$ we have $x_o \cdot c \in \mathfrak{a} \setminus \{0\}$ and thus $R \cdot x_o \cdot c = \mathfrak{a}$ for all $x_o \neq 0$. Hence we can apply Proposition 2.13 as in the last proof, only this time with $p(x_o, x_1) = x_1 \cdot x_o \cdot c$. ●

Both of the preceding results have an immediate consequence which will be a cornerstone in the investigation of quasi-compactifiable artinian rings, the discussion of which we now pursue.

4.8. Corollary. A quasi-compactifiable semisimple artinian ring R is finite.

proof: By Wedderburn's Theorem R is the finite sum of matrix rings over skew fields; as each of the underlying skew fields is then also quasi-compactifiable (via the canonical embeddings into R), these are all finite by Proposition 6 (or 7). Hence R too is finite. ●

4.9. Proposition. Let D be an infinite skew field and R a D-algebra. If R is quasi-compactifiable as a ring then R is a zero ring (i.e., $R^2 = (0)$).

proof: Suppose $r \cdot s \neq 0$ for some $r, s \in R$. Then for every $x_o = d \cdot r \in D \cdot r$ with $d \cdot r \neq 0$ there is an element x_1 of R (namely $d^{-1} \cdot s$) such that $x_1 \cdot x_o = r \cdot s$. Thus Proposition 2.13

with $d(x,y) = x - y$, $c = r \cdot s$, $S = T = D \cdot r \setminus \{0\}$ and $p(x_0,x_1) = x_1 \cdot x_0$ gives the desired contradiction. ●

It is well-known that a torsion-free artinian ring is an algebra over $\mathbb{Q}$ and possesses a left identity element. Thus Proposition 9 gives us

<u>4.10. Corollary</u>. The only quasi-compactifiable torsion-free artinian ring is (0) .

We next settle quasi-compactifiability in unital artinian rings. To do this we first need information on what happens when passing to homomorphic images.

<u>4.11. Lemma</u>. Let R and S be rings such that R is noetherian, and let $\mathfrak{a}$ be an ideal of S . Then $S \in \mathbf{c}(R)$ implies $S/\mathfrak{a} \in \mathbf{c}(R/R \cap \mathfrak{a})$.

<u>proof</u>: Set $\mathfrak{a}' = \mathfrak{a} \cap R$ and assume that $S \in \mathbf{c}(R)$. Let Σ be a set of polynomial equations over $R/\mathfrak{a}'$ finitely solvable in $R/\mathfrak{a}'$. W.l.o.g. Σ can be assumed to be of the form

$$\{\phi_i = 0;\ i \in I\} ,$$

where ϕ_i is a polynomial with constants in $R/\mathfrak{a}'$. "Lift" each ϕ_i to ϕ_i' by replacing all constants by representatives in R . Since R is noetherian there exist $a_1,\cdots,a_n$ elements of $\mathfrak{a}'$ such that $\mathfrak{a}' = Ra_1+\cdots+Ra_n$. Then the system of equations

$$\Sigma' = \{\phi_i' = z_i;\ i \in I\} \cup \{z_i = z_{i1}a_1+\cdots+z_{in}a_n;\ i \in I\}$$

(where the z_i's and z_{jk}'s are assumed to be variables not occurring in Σ), has constants in R and is finitely solvable in R (just "lift" solutions of members of Σ). Hence Σ' is solvable in S. But the z_i's are forced to take on values in $\mathfrak{a}$, and thus any solution of Σ' taken modulo $\mathfrak{a}$ yields a solution of Σ in $S/\mathfrak{a}$. •

<u>4.12. Proposition</u>. A quasi-compactifiable artinian ring R with identity is finite.

<u>proof</u>: By Proposition 2.12 we can choose $S \in \mathbf{c}(R)$ satisfying every positive sentence holding in R. We show first that $R/J(R)$ is quasi-compactifiable, and hence finite by Corollary 8. By the Akizuki-Hopkins Theorem R is also noetherian, hence this will be accomplished by Lemma 11 provided an ideal $\mathfrak{a}$ of S can be found satisfying $R \cap \mathfrak{a} = J(R)$. Now $J(R) = Ra_1 + \cdots$ $\cdots + Ra_n$ for suitable $a_1, \cdots, a_n$ because R is noetherian. We set

$$\mathfrak{a} = Sa_1 + \cdots + Sa_n$$

and claim that $\mathfrak{a}$ does the job. Now the two-sidedness of the left ideal $Ra_1 + \cdots + Ra_n$ is expressed by the positive sentence

$$\phi \equiv (\forall x_1) \cdots (\forall x_n)(\forall y)(\exists z_1) \cdots (\exists z_n) \\ ((x_1a_1 + \cdots + x_na_n) \cdot y = z_1a_1 + \cdots + z_na_n) .$$

Therefore ϕ must hold in S, which implies that the left ideal $\mathfrak{a}$ is also two-sided. Now obviously $\mathfrak{a} \cap R$ contains $J(R)$. To show the other inclusion recall that the artinian ring R has $J(R)$ as its largest nilpotent ideal; suppose that $J(R)^m = (0)$. This implies that the following positive sentence holds in R :

$$\theta \equiv (\forall x_{ij})_{i=1,\cdot\cdot,n,\ j=1,\cdot\cdot,m}\left(\prod_j \left(\sum_i x_{ij}a_i\right) = 0\right) .$$

But then Θ must hold in S , which in turn implies the nilpotency of $\mathfrak{a}$; hence $\mathfrak{a} \cap R$ is a two-sided nilpotent ideal of R and therefore contained in $J(R)$, which proves our claim. Now for arbitrary $n \in \mathbb{N}$, $J(R)^n/J(R)^{n+1}$ is a finitely generated (left) R-module, as R itself is noetherian, and therefore $J(R)^n/J(R)^{n+1}$ is a finitely generated $R/J(R)$-module since $J(R)$ is in the annihilator. Since $R/J(R)$ is finite, it follows that $J(R)^n/J(R)^{n+1}$ is finite for every $n \in \mathbb{N}$ and therefore R is finite as $J(R)$ is nilpotent. •

We are ready to prove :

4.13. Theorem. The following are equivalent conditions on an artinian ring R :

(i) $R^+ \cong B \oplus P$, where B is a finite group and P is a finite sum of Prüfer groups lying in $\mathrm{Ann}(R)$.

(ii) R is a retract of a compact ring.

(iii) R is equationally compact.

(iv) R is quasi-compactifiable.

(Observe that (ii) $\Leftrightarrow$ (iii) gives a positive answer to the Mycielski Question in the class of artinian rings and (iii) $\Leftrightarrow$ (iv) gives for this class a strong positive answer to the Weglorz Problem.)

proof: (i) $\Rightarrow$ (ii): If P is the sum of n Prüfer groups then by injectivity there is a retraction $h : C^n \to P$, where C is the (compact!) circle group; extending via the identity map on B we obtain a group retraction f of the compact group $H := B \oplus C^n$ onto R^+ . Now turn H into a ring $\widetilde{H}$ by retaining the multiplication defined on R^+ , letting

C^n annihilate $\tilde{H}$, and extending by distributivity. That f is a ring homomorphism is easily checked. Now the inverse image under the multiplication map of any subset of H is the finite union of pairs of cosets of C^n in H, all of which are closed in the product topology; thus multiplication is continuous, and $\tilde{H}$ is a compact ring retracting onto R.

(ii)$\Rightarrow$(iii): by Proposition 2.2.

(iii)$\Rightarrow$(iv) is trivial.

(iv)$\Rightarrow$(i): If R^+ were not a torsion group there would exist by [44,Satz 4] a proper torsion-free direct summand of R, and hence a proper torsion-free artinian ring quasi-compactifiable too, contradicting Corollary 10. Thus R^+ is torsion, and if $R^+ \cong B \oplus P$ is a representation of R^+ as the sum of reduced and divisible parts, then P is a direct sum of Prüfer groups (see [29]). Moreover $P \subseteq \mathrm{Ann}(R)$ by Proposition 3.7; thus the Prüfer group summands are ideals, and so by D.C.C. P is a finite sum of Prüfer groups.

It remains to show the finiteness of B. Let $\overline{B}$ be the subring of R generated by B. If $\mathfrak{a}$ is an arbitrary left ideal of $\overline{B}$ then $\mathfrak{a}$ is also a left ideal of R, because

$$R\cdot\mathfrak{a} = (P+B)\cdot\mathfrak{a} = P\cdot\mathfrak{a} + B\cdot\mathfrak{a} \subseteq (0) + \overline{B}\cdot\mathfrak{a} \subseteq \mathfrak{a}.$$

Thus $\overline{B}$ inherits the descending chain condition from R, i.e., $\overline{B}$ is artinian and, of course, quasi-compactifiable. We claim that $\overline{B}^+$ is a bounded torsion group. Since $\overline{B}$ is artinian the family of ideals $\{m\cdot\overline{B};\ m \in \mathbb{N}\}$ has a smallest element, say $n\cdot\overline{B}$, which is of course additively a divisible subgroup of R^+. Hence $n\cdot\overline{B} \subseteq P$. Since $n\cdot B \subseteq B$ we obtain

$$n \cdot B \subseteq n \cdot \overline{B} \cap B \subseteq P \cap B = \{0\} ,$$

and thus B is bounded torsion. That the underlying group of $\overline{B}$, the ring generated by B, is also bounded torsion, is then elementary. Thus Proposition 3.10 applies and $Z_n * \overline{B}$ is quasi-compactifiable. But $Z_n * \overline{B}$ is artinian because the $Z_n * \overline{B}$-modules $\overline{B}$ and $Z_n * \overline{B}/\overline{B}$ are artinian; thus $Z_n * \overline{B}$ is finite by Proposition 12 and so, of course, is B. •

<u>4.14. Corollary</u>. A compact artinian ring R is finite.

<u>proof</u>: Suppose we could endow the group $R^+ = B \oplus P_1 \cdots \oplus P_n$, where B is finite and $P_i = \mathbb{Z}(p_i^\infty)$, $i=1,\cdots,n$, with a compact topology. If P_i^k is the subgroup of P_i consisting of all complex p_i^k-th roots of unity, then R is the union of all the finite sets

$$R^k := B \oplus P_1^k \oplus \cdots \oplus P_n^k ,$$

$k \in \mathbb{N}$. By the Baire Category Theorem one of the sets $R \setminus R^k$ is not dense in R, so some R^k must contain a non-empty open set, forcing the topology to be discrete which, under the assumption that P_i's occur at all, is a contradiction. •

<u>4.15. Example</u>. If multiplication is defined on the group $Z_2 \oplus \mathbb{Z}(2^\infty)$ by letting $\mathbb{Z}(2^\infty)$ annihilate everything and setting $(1,0) \cdot (1,0) = (0,-1)$ ($\mathbb{Z}(2^\infty)$ written additively) we obtain a ring R demonstrating that the decomposition given in Theorem 13 (i) cannot, in general, be made ring direct; here a product of non-divisible elements is divisible. Note that R is even subdirectly irreducible.

Let us in conclusion point out some open questions and put the foregoing discussion in perspective with the results in the literature on compact topological rings. I. Kaplansky proved that a compact simple ring is finite [29, Thm. 9, Corollary]. The question left blatantly open is: are all equationally compact simple rings finite? Or stronger yet: are all quasi-compactifiable simple rings finite? We have however been able to sharpen Kaplansky's result under an additional hypothesis: an infinite simple ring with socle cannot even be <u>embedded</u> into a compact ring. For simple rings there are two natural subcases, the radical rings on the one hand and the semisimple rings on the other. All of the (very few!) known radical simple rings are not equationally compact. The semisimple side of the board means concentrating to begin with on primitivity. We showed in [20] that an equationally compact primitive ring with socle is finite, but we do not know if this holds under the weaker hypothesis of quasi-compactifiability. Again it was Kaplansky who proved that a compact primitive ring is finite [28, Theorem 16, Corollary], but whether an infinite primitive ring can be embedded into a compact ring is open. The standard example of a simple ring without socle - obtained by dividing the socle out of the ring of linear transformations of a countably infinite dimensional vector space - is easily seen to be not quasi-compactifiable (apply Proposition 2.13 with $c = 1$ and $p(x_o,x_1,x_2) = x_1 \cdot x_o \cdot x_2$). In Chapter III we shall see that an equationally compact simple (even subdirectly irreducible!) ring with A.C.C. is finite. With regard to our conclusive results on artinian rings it was, we believe, S. Warner who first showed that compact artinian rings with identity are finite [52, Theorem 2, Corollary]. We have sharpened this result in two directions:

first, by showing that the hypothesis demanding an identity is superfluous, and second, by showing that an infinite artinian ring with identity cannot even be embedded into a compact ring (Proposition 12). And those artinian rings which are embeddable into compact rings we have structurally characterized in Theorem 13, as this property lies in strength between conditions (ii) and (iv) given there.

CHAPTER III.
RINGS WITH THE ASCENDING CHAIN CONDITION

In [52] S. Warner poses the following question: for which classes of rings does each compact ring therein possess a <u>unique</u> compact topology? There it is shown that the class of (unital) noetherian rings has this property, the topology in question being the Jacobson radical topology; in another paper [51] he showed that the class of semisimple rings enjoys the same property, the proof of this latter result drawing heavily on I. Kaplansky's structure theorem for compact semi-simple rings; a simplified and independent proof of this we give in Chapter IV. In particular a compact topological ring with no non-zero topological nilpotents is in possession of the only compact topology it can carry. Note that Corollary 4.14 implies trivially that the class of artinian rings enjoys this property too.

In [18] we showed that in the class of commutative noetherian rings equational compactness is equivalent to compactness, so a strong motivating factor prompting a further investigation in this direction was the suspicion that the sharpened result obtained in the commutative noetherian case could be substantially generalized - yielding, first, an answer to the Mycielski Question in this larger class of rings and, secondly, providing additional <u>algebraic</u> footholds for a further assault on the uniqueness question. In this chapter positive answers to both questions are obtained for the class of rings with A.C.C.

§5 The noetherian and torsion cases

Our point of departure in the succeeding deliberations is

5.1. Proposition (S. Warner [52]). Let $(R, \mathcal{T})$ be a topological noetherian ring. Then $(R, \mathcal{T})$ is a compact topological ring if and only if $\mathcal{T}$ is the $J(R)$-topology, R is complete in it, $\bigcap(J(R)^n; n \in \mathbb{N}) = (0)$, and $R/J(R)$ is finite.

Our efforts to extend the above result to the non-unital case within the framework of the ring-topological argumentation given in [52] were in vain. We shall see below how the model theoretic aspects of equational compactness lend a helping hand in the resolution of the problem. In this section we derive the promised results for the noetherian and torsion cases, for we need these stepping stones in order to approach the general situation in the next paragraph.

5.2. Lemma. Let R be a subring of an equationally compact ring S, and let $\mathcal{T}$ be a not necessarily Hausdorff topology on R with respect to which the operations are continuous and which possesses a subbase of neighborhoods of zero consisting of left ideals of S, all of which have finite index in R and are closed in the structure topology $\mathcal{S}(S)$. Then $\mathcal{T}$ is quasi-compact.

proof: Let $\{\mathfrak{a}_i; i \in I\}$ be the given subbase. Then the family

$$\mathcal{F} = \{r + \mathfrak{a}_i ; i \in I, r \in R\}$$

is a subbase of open sets for $\mathcal{T}$. Since each $\mathfrak{a}_i$ has finite index in R, $\mathcal{F}$ is also a subbase of closed sets for $\mathcal{T}$. By the Alexander Subbase Theorem $\mathcal{T}$ is quasi-compact if every subset of $\mathcal{F}$ satisfying the finite intersection property has a non-empty intersection. So let

$$\mathcal{F}_o = \{r_j + \mathfrak{a}_{i_j} \ ; \ j \in J\}$$

be such a subset. Setting

$$\Sigma = \{x = r_j + x_j \ ; \ j \in J\} \cup \{x_j \in \mathfrak{a}_{i_j} \ ; \ j \in J\} ,$$

it is clear that the finite intersection property for $\mathcal{F}_o$ just means that Σ is finitely solvable in S. Since S is equationally compact, Proposition 2.7 guarantees the solvability of Σ in S. But the substitute for x in any solution of Σ lies in the meet of $\mathcal{F}_o$. ●

5.3. Lemma. Let R be a ring such that either R has an identity or R^+ is torsion. Let $\mathfrak{a}$ and $\mathfrak{b}$ be left ideals of R of finite index, and let $\mathfrak{b}$ be finitely generated. Then $\mathfrak{a}\cdot\mathfrak{b}$ has finite index.

proof: Since $R/\mathfrak{b}$ is finite it suffices to show that $\mathfrak{b}/\mathfrak{a}\cdot\mathfrak{b}$ is finite. Let $b_1,\cdots,b_k$ generate $\mathfrak{b}$ as a left ideal. If R^+ is torsion, let m be a natural number at least as large as each of the orders of $b_1,\cdots,b_k$. Otherwise, set $m = 1$. It follows that

$$\mathfrak{b} = \bigcup(Rb_1+\cdots+Rb_k + b_n \ ; \\ b_n = n(1)b_1+\cdots+n(k)b_k, \ n \in \{0,\cdots,m-1\}^{\{1,\cdots,k\}}) .$$

Further let $R = \bigcup(r_i + \mathfrak{a} \ ; \ i = 1,\cdots,l)$ be a finite covering of R by cosets of $\mathfrak{a}$, and let $b \in \mathfrak{b}$ be arbitrary.

Then

$$b = \sum_{j=1}^{k} s_j \cdot b_j + b_n$$

for appropriate $s_j \in R$ and $n \in \{0,\cdots,m-1\}^{\{1,\cdots,k\}}$. For each $j = 1,\cdots,k$ there exists $i_j \in \{1,\cdots,l\}$ such that $s_j \in r_{i_j} + \mathfrak{a}$ - say, $s_j = r_{i_j} + a_j$ with $a_j \in \mathfrak{a}$. Then

$$b = \sum_{j=1}^{k} s_j \cdot b_j + b_n = \sum_{j=1}^{k} (r_{i_j} + a_j)\cdot b_j + b_n =$$

$$= \sum_{j=1}^{k} r_{i_j} \cdot b_j + b_n + \sum_{j=1}^{k} a_j \cdot b_j \in \sum_{j=1}^{k} r_{i_j} \cdot b_j + b_n + \mathfrak{a}\cdot\mathfrak{b} .$$

Since there are only a finite number of r_i's, b_j's and b_n's, it is now clear that in $\mathfrak{b}$ there can be only a finite number of cosets modulo $\mathfrak{a}\cdot\mathfrak{b}$. •

5.4. Lemma. Let R be an equationally compact ring and let $\mathfrak{a}$ and $\mathfrak{b}$ be closed ideals (resp. left ideals) with $\mathfrak{a} \supseteq \mathfrak{b}$. Then $\mathfrak{a}/\mathfrak{b}$ is an equationally compact ring (resp. R-module).

proof: Suppose $\mathfrak{a}$ and $\mathfrak{b}$ are ideals and let Σ be a system of ring polynomial equations with constants in $\mathfrak{a}/\mathfrak{b}$, and which is finitely solvable in $\mathfrak{a}/\mathfrak{b}$. Σ can be assumed to be of the form

$$\Sigma = \{\phi_j = 0;\ j \in J\}$$

where each ϕ_j is a ring polynomial with constants in $\mathfrak{a}/\mathfrak{b}$. Replace each constant from $\mathfrak{a}/\mathfrak{b}$ which appears in ϕ_j by an arbitrary element from $\mathfrak{a}$ which represents it modulo $\mathfrak{b}$. The result is a ring polynomial ϕ_j' with constants in R, and it follows that

$$\Sigma' := \{\phi_j' = z_j;\ j \in J\} \cup \{z_j \in \mathfrak{b};\ j \in J\} \cup \{x_i \in \mathfrak{a};\ i \in I\}$$

(where z_j, $j \in J$, are variables not occurring in Σ, and x_i, $i \in I$, are all the variables occurring in Σ) is finitely solvable in R: for a finite subset of Σ', choose a solution of the corresponding subset of Σ, pick representatives of this solution out of the respective cosets and compute for the z_j's, which of necessity must lie in $\mathfrak{b}$. Thus Σ' is solvable in R because R is equationally compact and $\mathfrak{a}$ and $\mathfrak{b}$ are closed (Proposition 2.7). Any solution taken modulo $\mathfrak{b}$ yields a solution of Σ in $\mathfrak{a}/\mathfrak{b}$. The other statement is proven by appropriately modifying the above. •

Focussing now on the noetherian case, let us first record

5.5. Proposition. An equationally compact semisimple noetherian ring R is finite.

proof: R is linearly compact in the discrete topology, i.e., families of cosets of left ideals with the finite intersection property have a non-empty intersection. This follows from the fact that left ideals in noetherian rings (and hence cosets of same) are closed by Proposition 3.2, and therefore Proposition 2.7 applies. Thus by D. Zelinsky's structure theorem for semisimple linearly compact rings [61, Theorem 1] R is a product of simple artinian rings. As R is noetherian, there can only be a finite number of factors, and as each factor is finite (Corollary 4.8), R is too. •

It is not difficult to give a proof of the above independent of Zelinsky's structure theorem on linearly compact rings; see [19].

5.6. Lemma. Let R be an equationally compact noetherian ring, and let $\mathfrak{a}$ be a left ideal of R such that the R-module $R/\mathfrak{a}$ is subdirectly irreducible. Then $R/\mathfrak{a}$ is finite.

proof: Suppose not. Since $R/\mathfrak{a}$ is subdirectly irreducible there is an element $a \notin \mathfrak{a}$ contained in each left ideal properly containing $\mathfrak{a}$. In particular, for any $s \in R \setminus \mathfrak{a}$, the left ideal $\mathfrak{a} + Rs$ contains a. Choose a set I such that $|I| > |R|$, and let

$$\Sigma = \{a - y_{ij}(x_i - x_j) = z_{ij};\ i,j \in I,\ i \neq j\} \cup \{z_{ij} \in \mathfrak{a};\ i,j \in I,\ i \neq j\}.$$

Σ is finitely solvable: indeed, in any finite subset of Σ replace the occurring x_i's by elements of R which lie pairwise in distinct cosets of $\mathfrak{a}$, and then solve. $\mathfrak{a}$ is finitely generated, hence closed (Proposition 3.2), so Proposition 2.7 says that Σ is solvable in R. But the cardinality of I forces two distinct x_i and x_j to assume the same value, implying that $a \in \mathfrak{a}$, a contradiction. •

Now let R be an arbitrary ring with identity. By Birkhoff's Theorem there are subdirectly irreducible (left) R-modules M_i, $i \in I$, and an R-module embedding

$$R \rightarrowtail \prod(M_i;\ i \in I)$$

which is subdirect. Then the R-module $M := \bigoplus(M_i;\ i \in I)$ is faithful. Indeed, suppose $r \in R$ annihilates M. If $\pi_i : M \twoheadrightarrow M_i$ is the natural projection given by the subdirect representation above, then $\pi_i(r) = \pi_i(r\cdot 1) = r\cdot(\pi_i(1)) = 0$, since $\pi_i(1) \in M_i \subseteq M$. But this holds for each $i \in I$, which forces $r = 0$. Now endow R with the M-topology $\mathcal{T}$ - for $\mathcal{T}$, a neighborhood base of $r \in R$ is given by the sets

$$O_{r;m_1,\cdots,m_k} = \{s \in R;\ sm_i = rm_i,\ i=1,\cdots,k\},\quad k \in \mathbb{N},\ m_1,\cdots,m_k \in M\ .$$

It is well-known that $\mathcal{T}$ is a topology for R, and, in our case, $\mathcal{T}$ is Hausdorff since M is faithful. Moreover, a subbase of neighborhoods of zero is given by the family of annihilators $\mathrm{Ann}(m)$, $m \in M$. However, by the nature of M each $\mathrm{Ann}(m)$ is the finite intersection of the annihilators $\mathrm{Ann}(m_j)$ where $m_j \in M_{i_j}$ and $m = \sum m_j \in \bigoplus M_{i_j}$. Hence the family

$$\{\mathrm{Ann}(m);\ 0 \neq m \in M_i,\ i \in I\}$$

is a subbase of neighborhoods of zero. Now for any $i \in I$ and any non-zero $m \in M_i$, the R-module $R/\mathrm{Ann}(m)$ is isomorphic to $R \cdot m$. The latter is a proper submodule of the subdirectly irreducible R-module M_i, hence is itself subdirectly irreducible. Collecting all the pieces we see that there is a subbase of neighborhoods of zero for the topology $\mathcal{T}$ consisting of left ideals $\mathfrak{a}_j$, $j \in J$, such that $R/\mathfrak{a}_j$ is a subdirectly irreducible R-module. (This observation is implicit in the proof of [37, Lemma 3].)

5.7. Proposition. If R is an equationally compact noetherian ring, then the M-topology $\mathcal{T}$ (as constructed above) is a compact topology for R.

proof: The above remarks together with Lemma 6 yield that $\mathcal{T}$ is a Hausdorff topology on R with a subbase for zero consisting of left ideals each of finite index in R. Since in a noetherian ring every left ideal is closed (Proposition 3.2), Lemma 2 (putting $R = S$ there) then settles the matter. ●

An immediate consequence of Propositions 1 and 7 is

5.8. Corollary. A noetherian ring R is equationally compact if and only if the $J(R)$-topology is compact.

In the following two lemmata it is assumed that R is a ring satisfying A.C.C. and having a torsion group underlying it (which, under A.C.C. is equivalent to R having positive characteristic). Let $\chi(R) = n > 0$.

5.9. Lemma. If R is semisimple and equationally compact, then R is finite.

proof: By Proposition 3.10 $Z_n * R =: S$ is an equationally compact (noetherian!) ring. We show that S is finite. Now $J(S)$ is closed in S so by Lemma 4 $S/J(S)$ is equationally compact; moreover $S/J(S)$ is noetherian and semisimple, hence finite by Proposition 5. Thus it would suffice to know that $J(S)$ is finite. Suppose not; then by the finiteness of Z_n there would be elements $r \neq s$ in R and $z \in Z_n$ such that $r + z$ and $s + z$ are elements of $J(S)$. But then, remembering that R is an ideal in S,

$$0 \neq r - s = (r + z) - (s + z) \in J(S) \cap R = J(R),$$

a contradiction. ●

5.10. Lemma. If $\mathcal{T}$ is an arbitrary compact topology for R, then any left ideal $\mathfrak{a}$ of finite index is open in $\mathcal{T}$.

proof: Let $\mathfrak{a}$ be generated by $a_1, \cdots, a_m$. Then $Ra_1 + \cdots$ $\cdots + Ra_m$ is the image of the compact space R^m under the continuous map $(x_1,\cdots,x_m) \mapsto x_1a_1+\cdots+x_ma_m$, hence is compact, and thus closed. But then

$$\mathfrak{a} = \bigcup(Ra_1+\cdots+Ra_m+z_1a_1+\cdots+z_ma_m;\ 0 \le z_i < n,\ i=1,\cdots,m)$$

is also closed, hence open because of its finite index. ●

5.11. Proposition. Let R be an equationally compact ring of positive characteristic n and satisfying A.C.C. Then R is a compact ring and the $J(R)$-topology is the unique compact topology for R.

proof: $S = Z_n * R$ is equationally compact by Proposition 3.10 and also noetherian. By Corollary 8 the $J(S)$-topology is compact. Since $J(R) \subseteq J(S)$ it follows that

$$\bigcap(J(R)^n;\ n \in \mathbb{N}) \subseteq \bigcap(J(S)^n;\ n \in \mathbb{N}) = (0),$$

i.e., the $J(R)$-topology is Hausdorff. Now $J(R)$ and all its powers are closed in R by Proposition 3.2. Thus $R/J(R)$ is equationally compact by Proposition 2.8, and therefore satisfies the hypotheses of Lemma 9. Hence $R/J(R)$ is finite. Calling on Lemma 3 a simple induction argument yields that $J(R)^n$ has finite index for each $n \in \mathbb{N}$. The $J(R)$-topology thus satisfies the hypotheses of Lemma 2 (again putting $R = S$ there) and we conclude that it is a compact topology for R. Since it possesses a subbase of neighborhoods for zero consisting of left ideals of finite index, it must be coarser that any other compact topology for R by Lemma 10. Since distinct compact topologies are never comparable, the $J(R)$-topology is the unique compact topology for R. ●

§6 The characterization theorem

In this section R will always denote a ring satisfying A.C.C. The analysis of the J(R)-topology by a frontal attack, as was possible in the torsion case, eludes us here. We no longer have Lemma 5.3 at our disposal, nor is there any reason to believe that the powers of the radical are closed ideals, tools which found heavy use in the last section. We must take a different tack, by constructing first a more amenable topology for R - later it will reveal itself in the presence of equational compactness as just a different discription of the J(R)-topology.

6.1. Proposition. Let R be equationally compact. Then

$$R \cong R_1 \oplus \cdots \oplus R_n$$

where, for each $i = 1,\cdots,n$, R_i is a $\mathbb{Z}^*_{p_i}$-algebra, complete in its Hausdorff p_i-adic topology.

proof: Suppose $D \neq (0)$ is a divisible subgroup of R^+. Then D contains as a subgroup either $\mathbb{Q}^+$ or a Prüfer group $\mathbb{Z}(p^\infty)$ for some $p \in \mathbb{P}$ (see [29]). Both of these groups have a proper ascending chain of subgroups; since $D \subseteq \mathrm{Ann}(R)$ by Proposition 3.7 the subgroups in these chains are ideals in R, contradicting A.C.C. Hence R^+ is reduced and therefore by Proposition 3.5 $R \cong \prod(R_p;\ p \in \mathbb{P})$ with the properties described there. But almost all of these R_p's must be zero, again by A.C.C. ●

6.2. Lemma. Let R be equationally compact and subdirectly irreducible. Then R is finite.

proof: Subdirect irreducibility and Proposition 1 imply that $R \cong R_{p_i}$ (for some i) with the properties stated there. Set $p = p_i$. In particular, then, $\bigcap(p^n R;\ n \in \mathbb{N}) = (0)$; since the $p^n R$'s are ideals some $p^n R$ must be (0) and therefore $\chi(R) = p^n$. By Proposition 5.11 R is a compact ring in which the unique compact topology is the $J(R)$-topology. In particular, $\bigcap(J(R)^n;\ n \in \mathbb{N}) = (0)$. Again subdirect irreducibility implies $J(R)^n = (0)$ for some n , that is, the $J(R)$-topology is discrete. Being compact also means that R has to be finite, which was the claim. •

Let R be equationally compact. From Proposition 1 it follows that

$$\bigcap(mR;\ m \in \mathbb{N}) = (0)\ .$$

Let $0 \neq r \in R$. Then there is an $m \in \mathbb{N}$ such that $r \notin mR$. Zorn's Lemma yields an ideal, say $\mathfrak{a}_r$, which is maximal with respect to the property of (1) being an ideal, (2) not containing r , and (3) containing mR . Let $\mathcal{T}$ denote the topology on R generated by

$$\{ a + \mathfrak{a}_r\ ;\ a \in R\ ,\ r \in R \setminus \{0\} \}$$

taken as a subbase for the open sets. Being an ideal topology $\mathcal{T}$ is a topology for R , and moreover is Hausdorff since $\bigcap(\mathfrak{a}_r\ ;\ r \in R \setminus \{0\}) = (0)$. The following is easily verified:

6.3. Lemma. Let R be equationally compact. Then for $0 \neq r \in R$, $\mathfrak{a}_r$ is a closed ideal, being the solution set of the positive formula

$$(\exists x_1)\cdots(\exists x_{t+1}) \bigvee (x_o = x_1a_1+\cdots+x_ta_t+ +z_1a_1+\cdots+z_ta_t+ mx_{t+1};\ z_i \in \mathbb{N},\ 0 \le z_i < m),$$

where $a_1,\cdots,a_t$ generate $\mathfrak{a}_r$ as left ideal and $m \in \mathbb{N}$ is such that $\mathfrak{a}_r \supseteq mR$.

6.4. Lemma. Let R be equationally compact. Then $\mathcal{T}$ is a compact topology.

proof: It is clear that $\mathfrak{a}_r$ is maximal with respect to the property of being an ideal not containing r . Thus $R/\mathfrak{a}_r$ is subdirectly irreducible, satisfies A.C.C., and is equationally compact since $\mathfrak{a}_r$ is closed. Hence $R/\mathfrak{a}_r$ is finite by Lemma 2, i.e., $\mathfrak{a}_r$ has finite index. Then Lemma 5.2 applies (taking R = S again), implying that $\mathcal{T}$ is quasi-compact, and hence compact. •

We are ready to prove:

6.5. Theorem. The following are equivalent conditions on a ring R satisfying the ascending chain condition on left ideals:

(i) R is a compact ring.

(ii) R is equationally compact.

(iii) R is a subdirect product of a family of finite subdirectly irreducible rings, closed with respect to the product of the discrete topologies on the factors.

(iv) R is an ideal of an equationally compact noetherian ring.

proof: (i)$\Leftrightarrow$(ii) follows from Lemma 4.

(iii)$\Rightarrow$(i) is clear since the product topology is compact.

(iv)$\Rightarrow$(ii): Ideals are closed in noetherian rings, so Corollary 2.10 applies.

(ii)$\Rightarrow$(iii): For R we have the topology $\mathcal{T}$ as constructed above. Set $R^\wedge = R \setminus \{0\}$. The canonical projections of R onto $R/\mathfrak{a}_r$, $r \in R^\wedge$, induce an embedding

$$R \rightarrowtail \prod (R/\mathfrak{a}_r \; ; \; r \in R^\wedge) \quad =: \quad S$$

representing R as a subdirect product of the finite subdirectly irreducible rings $R/\mathfrak{a}_r$. To show is that R is closed in S, where S is endowed with the product of the discrete topologies on its factors. Identify R with its image in S and suppose that $\bar{a} := (\bar{a}_r)_{r \in R^\wedge}$, where $\bar{a}_r = a_r + \mathfrak{a}_r \in R/\mathfrak{a}_r$, lies in the closure of R. This means that every subbasic neighborhood of $\bar{a}$ meets R, i.e., that for every finite set $r_1, \cdots, r_n \in R^\wedge$ there exists $b \in R$ such that $b + \mathfrak{a}_{r_i} = \bar{a}_{r_i}$, i.e., such that $b - a_{r_i} \in \mathfrak{a}_{r_i}$, $i = 1, \cdots, n$. But that just means that the system

$$\Sigma = \{x - a_r = y_r \; ; \; r \in R^\wedge\} \cup \{y_r \in \mathfrak{a}_r \; ; \; r \in R^\wedge\}$$

is finitely solvable in R. Since the $\mathfrak{a}_r$'s are closed in R (Lemma 3) Σ is solvable in R by Proposition 2.7, yielding an element $a \in R$ (the substitute for x) with $a = (\bar{a}_r)_{r \in R^\wedge} = \bar{a}$, that is to say, $\bar{a} \in R$.

(ii)$\Rightarrow$(iv): Let $R = R_1 \oplus \cdots \oplus R_n$ be the decomposition of R given by Proposition 1. Since each R_i also has A.C.C. and is equationally compact, it suffices to prove the claim for each of the summands; that is, R may be assumed to be a $\mathbb{Z}_p^*$-algebra and complete in its Hausdorff p-adic topology.

Set $S = \mathbb{Z}_p^* * R$. S is noetherian, R is an ideal of S , so the claim will be proved if a compact topology for S can be found. Let $\mathcal{T}_S$ be the topology on the set $\mathbb{Z}_p^* \times R$ given by taking the product of the (compact!) p-adic topology on $\mathbb{Z}_p^*$ with the topology $\mathcal{T}$ on R , which is compact by Lemma 4. Thus $\mathcal{T}_S$ is compact, and we claim it is a topology for S . Now obviously $\mathcal{T}_S$ is compatible with the addition (being componentwise) and it is equally clear that the multiplication map, when restricted to either of the factors $\mathbb{Z}_p^*$ or R , is continuous. Thus the only point of concern is the continuity of the map

$$\mu : \mathbb{Z}_p^* \times R \longrightarrow R$$
$$(z,s) \longmapsto z \cdot s \ .$$

Now $z \cdot s + \mathfrak{a}_r$ is a typical subbasic neighborhood of $z \cdot s$. By construction $\mathfrak{a}_r$ contains mR for some $m \in \mathbb{N}$. Then for some $n \in \mathbb{N}$, $m\mathbb{Z}_p^* = p^n \mathbb{Z}_p^* = (p^n)$, a neighborhood of zero in the p-adic topology on $\mathbb{Z}_p^*$. Moreover, $\mathfrak{a}_r$ is a $\mathbb{Z}_p^*$-algebra ideal, since every element in $\mathbb{Z}_p^*$ is the sum of some integer and an element of (p^n) and

$$(p^n) \cdot \mathfrak{a}_r = p^n \cdot \mathbb{Z}_p^* \cdot \mathfrak{a}_r = m \cdot \mathbb{Z}_p^* \cdot \mathfrak{a}_r \subseteq m \cdot R \subseteq \mathfrak{a}_r \ .$$

Thus $(z + (p^n)) \times (s + \mathfrak{a}_r)$ is a neighborhood of (z,s) and an easy computation now shows that it is mapped by μ into $z \cdot s + \mathfrak{a}_r$. ●

Remark. Implicit in the proof of (ii) $\Rightarrow$ (iv) above is the following: If R is an equationally compact ring satisfying A.C.C. then (in the notation of Proposition 1) R is even a compact topological A-algebra, where $A = \mathbb{Z}_{p_1}^* \oplus \cdots \oplus \mathbb{Z}_{p_n}^*$ is endowed with its (unique) compact topology and R with its compact topology $\mathcal{T}$.

We return to the radical topology; now, condition (iv) of Theorem 5 offers a suitable setting in which to study it.

<u>6.6. Lemma</u>. If R is equationally compact then $J(R)^n$ has finite index for every $n \in \mathbb{N}$.

<u>proof</u>: $J(R)$ is closed by Proposition 3.1. Hence $\overline{R} := R/J(R)$ is equationally compact by Proposition 2.8, satisfies A.C.C., and is semisimple. Then by Theorem 5 (iv) $\overline{R}$ is an ideal of an equationally compact noetherian ring S. Then $S/J(S)$ is equationally compact, semisimple and noetherian, hence finite by Proposition 5.5. But $S/J(S)$ is at least as large as $\overline{R}/(J(S)\cap\overline{R})$; since $J(S)\cap\overline{R} = J(\overline{R}) = (0)$ we conclude that $\overline{R}$ is finite, i.e., that $J(R)$ has finite index in R. Thus the Lemma will be proved if we show the following: For each $n \geq 1$, $J(R)^n/J(R)^{n+1}$ is finite.

Now let S be an equationally compact noetherian ring containing R as an ideal (Theorem 5 (iv) again). $J(R)$ is then an ideal in S, being the meet of $J(S)$ and R, two ideals in S. Hence every power of $J(R)$ is an ideal in S, and therefore closed in S, S being noetherian; hence by Lemma 5.4 $J(R)^n/J(R)^{n+1}$ is an equationally compact ring. On the other hand $R/J(R)^{n+1}$ is a left R-module and has, as such, the ascending chain condition on submodules. This means that the R-module

$$M := J(R)^n/J(R)^{n+1}$$

is finitely generated over R ; but $J(R) \subseteq \mathrm{Ann}(M)$, and so M is a module over $R/J(R) = \overline{R}$ and finitely generated as such. Since $\overline{R}$ is finite, we conclude that M is even finitely generated as an abelian group. By the Fundamental Theorem, if M were infinite, M would contain a copy of $\mathbb{Z}^+$ as a (group) direct summand. But the group M is the abelian group underlying the equationally compact ring $J(R)^n/J(R)^{n+1}$, hence must be equationally compact as a group, and therefore the summand $\mathbb{Z}^+$ would be equationally compact. This is not the case, as $\mathbb{Z}^+$ is obviously not algebraically compact, and therefore M must be finite. •

<u>6.7. Proposition</u>. If R is equationally compact then the $J(R)$-topology is compact.

<u>proof</u>: By Theorem 5 (iv) R is an ideal of an equationally compact noetherian ring S . Since the $J(S)$-topology is compact, it follows as before that

$$\bigcap (J(R)^n;\ n \in \mathbb{N}) \subseteq \bigcap (J(S)^n;\ n \in \mathbb{N}) = (0) ,$$

i.e., the $J(R)$-topology is Hausdorff. As observed in the proof of Lemma 6, the powers of $J(R)$ are all closed in S ; this fact and Lemma 6 itself means that the conditions of Lemma 5.2 are met. We conclude that the $J(R)$-topology is compact. •

<u>6.8. Theorem</u>. If R is a compact ring satisfying A.C.C. then the $J(R)$-topology is the unique compact topology for R .

proof: In light of Proposition 7 it will suffice to show that the topology $\mathcal{T}$ is the coarsest compact topology for R. So let $\mathcal{T}_0$ be an arbitrary compact topology for R. To verify is that each subbasic $\mathcal{T}$-neighborhood of zero, $\mathfrak{a}_r$, is $\mathcal{T}_0$-open. Let $a_1,\cdots,a_t$ be left ideal generators of $\mathfrak{a}_r$ and let $m \in \mathbb{N}$ be such that $\mathfrak{a}_r \supseteq mR$. Then

$$A \ := \ Ra_1 + \cdots + Ra_t + mR$$

is the image of R^{t+1} (equipped with the $\mathcal{T}_0$ power topology) under the continuous map

$$(x_1,\cdots,x_{t+1}) \ \mapsto \ x_1a_1 + \cdots + x_ta_t + mx_{t+1} \ ;$$

hence A is compact, hence $\mathcal{T}_0$-closed in R. Thus

$$\mathfrak{a}_r \ = \ \bigcup(A + z_1a_1+\cdots+z_ta_t \ ; \ 0 \leqslant z_i < m, \ 1 \leqslant i \leqslant t)$$

is also $\mathcal{T}_0$-closed, hence $\mathcal{T}_0$-open since $\mathfrak{a}_r$ has finite index. ●

With the help of the above we obtain another characterization of equational compactness. The details missing for the proof are standard ring-topological arguments, which may be supplied by the reader.

<u>6.9. Theorem</u>. A ring R satisfying A.C.C. is equationally compact if and only if $R/J(R)^n$ is finite for each $n \in \mathbb{N}$, $\bigcap(J(R)^n;\ n \in \mathbb{N}) = (0)$, and R is complete in the $J(R)$-topology.

Note that in the unital or torsion case the condition that each power of $J(R)$ have finite index in the above can be replaced by the weaker condition that just $R/J(R)$ be finite - on account of Lemma 5.3. However, in the general case the finiteness of $R/J(R)$ will not suffice - the zero ring over the group $\mathbb{Z}^+$ provides a simple example.

Let us conclude the chapter with a few remarks on the cardinalities of equationally compact rings; there are no powerful general results in this direction except for the A.C.C. case, where however the main results of this chapter yield conclusive answers.

Are there countably infinite equationally compact rings? An affirmative answer is easily given. Taking the zero rings over the group $\mathbb{Q}^+$ of rationals and over $\mathbb{Z}(p^\infty)$ gives both torsion and torsion-free examples, the underlying groups being divisible. On the reduced side of the board, one can take the zero ring W over the group $\bigoplus(Z_p^+;\ n \in \mathbb{N})$ for fixed prime p ; W is countable and equationally compact because the underlying group is a linear space over the field Z_p and hence injective in the equational class of abelian groups. Sharpening our focus a bit, what about the same question restricted to rings with identity? Corollary 3.11 yields an example of such a ring, namely $Z_p * W$ with W given above. Of course no countably infinite rings can be compact, because there are not even any countably infinite compact abelian groups (A. Hulanicki [25]). $Z_p * W$ settles the torsion side, what about the characteristic zero case? Here the answer is negative:

<u>6.10. Proposition</u>. A ring R which is a compactification

of the ring $\mathbb{Z}$ is uncountable.

proof: Let $\pi : R \twoheadrightarrow R/D(R)$ denote the canonical projection. $R/D(R)$ is equationally compact by Corollary 3.8 and, of course, reduced. Moreover,

$$\pi|_{\mathbb{Z}} : \mathbb{Z} \longrightarrow R/D(R)$$

is an embedding, as otherwise $\mathbb{Z} \cap D(R) \neq (0)$ and so there would be a non-zero element of $\mathbb{Z}$ annihilating all of R (Proposition 3.7), which cannot be. Thus w.l.o.g. we may assume that R is reduced. But then we have the decomposition

$$R \cong \prod(R_p \ ; \ p \in \mathbb{P})$$

given by Proposition 3.5. There are now two possibilities:

case 1. an infinite number of the R_p's occurring are not (0) . But then $|R| \geq 2^{\aleph_0}$ which was the claim.

case 2. $R \cong R_{p_1} \oplus \cdots \oplus R_{p_n}$. As $\chi(R) = 0$ at least one of the R_{p_i}'s , call it R_p , has a non-discrete p-adic topology. Supposing that $|R| = \aleph_0$, then $|R_p| = \aleph_0$ too. Thus R_p is a countable complete metric space and the countable set $\{R_p \setminus \{a\} \ ; \ a \in R_p\}$ of open dense sets has an empty intersection, contradicting the Baire Category Theorem. ●

Remark. There does not appear to be a simpler proof of the above statement. Note that a lot of heavy labor is taken over by the results of Chapter II. We record also that both cases in the above proof actually occur : $\mathbb{Z}$ is canonically embedded into the product of fields $\prod(Z_p \ ; \ p \in \mathbb{P})$, which fits into case 1, and $\mathbb{Z} \subseteq \mathbb{Z}_p^*$ for any p , which illustrates case 2.

A well-known result in topology asserts that an uncountable compact metric space has the cardinality of the continuum. Moreover, the Baire Category argument given at the end of the last proof says that a countably infinite ring cannot carry a compact metric topology. Together with Proposition 7 this yields a very pretty result, with which we conclude.

6.11. Theorem. An infinite equationally compact ring with A.C.C. has the power of the continuum.

CHAPTER IV.
DISCRIMINATOR VARIETIES AND m-RINGS

This chapter is devoted to the investigation of compactness and injectivity in discriminator varieties, these being the natural universal algebraic "parents" of arithmetic varieties of rings, alias m-rings - that is, equational classes of rings satisfying an identity $x^m = x$. Arithmetic varieties of rings were characterized by G. Michler and R. Wille [36] and fall, by their result, under the above mentioned universal algebraically defined heading of discriminator varieties: equational classes generated by a class of algebras possessing the operation

$$t(x,y,z) = \begin{cases} z & \text{if } x = y \\ x & \text{if } x \neq y \end{cases}$$

as a polynomial. The importance of the "discriminator polynomial" t was first realized by A. Pixley, who, along with H. Werner, investigated the far-reaching consequences of this operation for algebras possessing it as a polynomial (resp. algebraic function) and for the structure of equational classes generated by such algebras (see A. Pixley [40], [41] and H. Werner [58]).

The most relevant outside source having a bearing on this chapter is the paper of S. Bulman-Fleming and H. Werner [8]. Indeed, our results in §8 are natural generalizations and improvements of their work on quasi-primal varieties. In §8 we answer positively their conjecture regarding the extension of their results to the case of "decent" discriminator varieties and settles also the "uniqueness of topology" question for the compact members. With regard to this latter result, Theorem 8.21, thanks are due to Heinrich Werner, with whose collaboration important points in the proof were clarified. The main result of §9 generalizes the corresponding result on quasi-primal varieties due to H. Werner [59].

We begin this chapter with a short paragraph on Boolean extensions centering around a result of S. Burris, for which we give an algebraic proof. This result is not only an integral part of the theorems in §8, but yields, as a by-product, a significant class of examples of equationally compact rings with identity which cannot be topologized with a compact topology.

§7 Boolean extensions

It will not be the intent of this section to go into a general discussion of the theory of Boolean extensions (we refer the reader to S. Burris [9] and the wealthy bibliography given there for a survey of this field), but rather to concern ourselves with a specific aspect pertinent to our investigations on discriminator varieties in §8.

The Boolean extension is a universal algebraic construction lying, in its generality, between the power construction and the subdirect power construction. (See e.g. M. Gould and G. Grätzer [15] for specifics on this point.) If $\mathfrak{A} = \langle A;F \rangle$ is an algebra and I is a set, consider the power algebra $\mathfrak{A}^I$. With every I-tuple $f \in A^I$ we can associate its "inverse" $f^{-1} : A \longrightarrow P(I)$ defined by

$$f^{-1}(a) = \{i \in I \ ; \ f(i) = a\} .$$

It is clear that the set of "inverses" f^{-1}, $f \in A^I$, are precisely those maps p from A into $P(I)$ which " A - partition I ", that is, which satisfy

(1)
(i) $p(a) \cap p(b) = \emptyset$, for $a,b \in A$ and $a \neq b$, and
(ii) $\bigcup(p(a) \ ; \ a \in A) = I$.

We denote the set of A - partitions of I by $A[P(I)]$; the map $e : f \mapsto f^{-1}$ is obviously a bijection between A^I and $A[P(I)]$. In order that e become an isomorphism of algebras, we must clearly define the algebraic structure on $A[P(I)]$ as follows: if $h \in F$ is an n-ary operation, then for $f_1^{-1}, \dots$ $\dots, f_n^{-1} \in A[P(I)]$ define $h(f_1^{-1},\dots,f_n^{-1}) \in A[P(I)]$ by

$$(2)\qquad h(f_1^{-1},\cdots,f_n^{-1})(a) = \bigcup(f_1^{-1}(a_1)\cap\cdots\cap f_n^{-1}(a_n);\ a = h(a_1,\cdot\cdot,a_n),\ a_1,\cdot\cdot,a_n \in A),$$

for $a \in A$.

Now consider, in place of the Boolean algebra $P(I) = \langle P(I); \cup, \cap, \emptyset, I, ' \rangle$, an arbitrary Boolean algebra $B = \langle B; \cup, \cap, 0, 1, ' \rangle$ (we shall in this section borrow the set theoretic operation symbols $\cap$ and $\cup$ to denote the abstract meet and join operations in B, for reasons which will become apparent). We define now the Boolean extension (or power) of $\mathfrak{A}$ by B to be the algebra $\mathfrak{A}[B] = \langle A[B]; F \rangle$, where $A[B]$ is the set of A - partitions of 1 in the sense of (1) with $\emptyset = 0$ and $I = 1$, and where the operations in F are defined as in (2). If $\mathfrak{A}$ is infinite, then B is assumed to be complete in order that the definition be meaningful. Note that, since every Boolean algebra is a subalgebra of a power set Boolean algebra, every Boolean extension is a subalgebra of some $\mathfrak{A}[P(I)]$, that is to say, is isomorphic to a subalgebra of some power $\mathfrak{A}^I$, and as such it is subdirect containing the diagonal. If $\mathfrak{A}$ is finite this subdirect power representation can be made more explicit: Consider B^*, the Stone space of B (B^* is the Boolean space, i.e., the totally disconnected compact Hausdorff space, possessing B as its Boolean algebra of clopen sets). Then it is clear that via the isomorphism e, $A[B]$ corresponds to the set of continuous maps from B^* into A, where A is endowed with the discrete topology. In this chapter it will be convenient to view a Boolean extension of a finite algebra $\mathfrak{A}$ in either fashion, sometimes as $\mathfrak{A}[B]$, as defined, and other times as the subdirect power of $\mathfrak{A}^{B^*}$ of continuous maps.

Our algebraic presentation of Burris' theorem requires the introduction of a further algebraic construction: let $\mathcal{B} = \langle B; F \rangle$ be an algebra possessing a one-element subalgebra (0) and let I be a non-empty set. For $i,j \in I$ define $p_{ij} : B^I \to B^I$ by

$$(p_{ij}(a))(k) = \begin{cases} a(i) & \text{if } k = j \\ 0 & \text{otherwise} \end{cases}$$

(p_{ij} projects the i-th component onto the j-th spot and sets 0's everywhere else). We call the algebra

$$\mathcal{B}_p^I = \langle B^I ; F \cup \{p_{ij} ; i,j \in I\} \rangle$$

a "direct power of $\mathcal{B}$ with projections".

<u>7.1. Proposition</u>. If $\mathcal{B}$ is an equationally compact algebra containing a one-element subalgebra (0) then every direct power of $\mathcal{B}$ with projections, $\mathcal{B}_p^I$, is equationally compact.

<u>proof</u>: One checks easily that the p_{ij}'s are endomorphisms of the algebra $\mathcal{B}^I$. Moreover, for each $i,j,k,l \in I$,

$$p_{ij} \circ p_{kl} = \delta_{il} p_{kj} ,$$

where δ_{il} is the Kronecker delta. We may therefore assume that an arbitrary polynomial equation over $\mathcal{B}_p^I$ is of the form

(3) $(p=q)(a_1,\cdots,a_l,x_1,\cdots,x_m, p_{i_1j_1}(y_1),\cdots,p_{i_nj_n}(y_n))$

where $a_1,\cdots,a_l \in B^I$, the x_k's and y_k's are variables, and

$$(p=q)(a_1,\cdots,a_l, x_1,\cdots,x_m, y_1,\cdots,y_n)$$

is a polynomial equation over $\mathcal{B}^I$. (The endomorphism property insures that the p_{ij}'s "factor through" all the operations from F .) Now associate with each such polynomial equation (3) the system

$$(4) \quad \Big\{(p=q)(a_1(j_k),\cdots,a_1(j_k),\ x_1^{(j_k)},\cdots,x_m^{(j_k)},\ y_1^{(i_k)},\cdots \cdots,y_n^{(i_k)};\ k=1,\cdots,n\Big\} \cup \Big\{(p=q)(a_1(i),\cdots,a_1(i),\ x_1^{(i)},\cdots,x_m^{(i)},\ 0,\cdots,0)\ ;\ i\in I\setminus\{j_1,\cdots,j_n\}\Big\}$$

of polynomial equations over $\mathcal{L}$. It is now easily verified that a system Σ of polynomial equations over $\mathcal{L}_p^I$, finitely solvable in $\mathcal{L}_p^I$, reduces via the transition (3) to (4) to a finitely solvable system over $\mathcal{L}$, which is then solvable in $\mathcal{L}$ as $\mathcal{L}$ is equationally compact; and a solution yields, via the identification $z = (z^{(i)})_{i\in I}$ for any variable z, a solution of Σ in $\mathcal{L}_p^I$. ●

With this in hand we can easily circumvent the model theoretic techniques using a translation of languages due to Ershov, upon which the original proof (see [9]) of the following result hinges:

<u>7.2. Theorem (S. Burris [9])</u>. Let $\mathfrak{A}$ be a finite algebra and B a complete Boolean algebra. Then $\mathfrak{A}[B]$ is equationally compact.

<u>proof</u>: By a result of B. Weglorz [53] (see 9.2) B is equationally compact; therefore B_p^A is equationally compact by Proposition 1. We will be finished via a direct application of Proposition 2.9 after verifying the following:

(i) $A[B]$ is a closed set in B_p^A.

(ii) $\mathfrak{A}[B]$ is a subreduct of B_p^A.

To (i): Fix $a\in A$ and let $1_a \in B^A$ be defined by

$$1_a(b) = \begin{cases} 1, & \text{if } a = b \\ 0, & \text{otherwise} . \end{cases}$$

Then $A[B]$ is the solution set of the following positive formula over B_p^A :

$$\bigwedge(p_{ca}(x) \cap p_{da}(x) = 0;\ c,d \in A,\ c \neq d) \wedge (\bigcup(p_{ca}(x);\ c \in A) = 1_a).$$

To (ii): Let $f \in F$ be an n-ary fundamental operation on $\mathfrak{A}[B]$, and let $u_1, \cdots, u_n \in A[B]$. Then for $a \in A$ compute:

$$f(u_1,\cdot\cdot,u_n)(a) \overset{(\text{def.})}{=} \bigcup(\bigcap(u_i(a_i);i=1,\cdot\cdot,n);\ a = f(a_1,\cdot\cdot,a_n))$$

$$= \bigcup(\bigcap(p_{a_i a}(u_i)(a);\ i = 1,\cdots,n);\ a = f(a_1,\cdots,a_n))$$

$$= (\bigcup(\bigcap(p_{a_i a}(u_i)\ ;\ i = 1,\cdots,n)\ ;\ a = f(a_1,\cdots,a_n)))(a) .$$

Now for $b \in B$ let $b_a \in B^A$ be defined by

$$b_a(c) = \begin{cases} b & \text{if } a = c \\ 0 & \text{otherwise.} \end{cases}$$

Then obviously $\hat{b} = \bigcup(\hat{b}(a)_a;\ a \in A)$ for $\hat{b} \in B^A$. It follows that $f(u_1,\cdots,u_n)$ is given by

$$\bigcup\{[(\bigcup(\bigcap(p_{a_i a}(u_i);\ i=1,\cdot\cdot,n);\ a = f(a_1,\cdot\cdot,a_n)))(a)]_a;\ a \in A\}$$

$$= \bigcup(\bigcap(p_{a_i a}(u_i);\ i=1,\cdots,n);\ a = f(a_1,\cdots,a_n)) ,$$

and the latter is a polynomial expression over B_p^A in the parameters $u_1,\cdots,u_n$. ●

We conclude this section with a concrete class of examples of equationally compact unital rings which are not compact:

7.3. Example. A "regular open" subset of a topological space is an open set which is identical with the interior of its closure. Let B be the set of regular open sets of the real interval $[0,1]$ endowed with the usual topology. Then, as is well-known, $B = \langle B; \inf, \sup, \emptyset, [0,1], ' \rangle$ is a complete Boolean algebra in which

$$\inf (M_i;\ i \in I) = (\bigcap (M_i;\ i \in I))^{\circ} ,$$

$$\sup (M_i;\ i \in I) = (\overline{\bigcup (M_i;\ i \in I)})^{\circ} , \quad \text{and}$$

$$M' = ([0,1] \setminus M)^{\circ} ,$$

where $^{-}$ and $^{\circ}$ denote the closure and interior operators, respectively.

Now let F be a finite field and define R to be the ring $F[B]$. We can picture R as a set of "fuzzy" $[0,1]$ - tuples over F :

$$R = \{ f \in F^{[0,1] \setminus N};\ N \subseteq [0,1],\ N^{\circ} = \emptyset,\ f^{-1}(a) \in B \text{ for all } a \in F \}$$

where the operations are first carried out componentwise where defined, and then the resulting inverse image of every $a \in F$ is completed using the operator $(^{-})^{\circ}$; that is, for $a \in A$, if $M_a \subseteq [0,1]$ is the resulting set of components upon which the value a is assumed, let $x \in [0,1]$ also assume the value a if $x \in \overline{M_a}^{\circ}$.

Now R is equationally compact by Theorem 2. We wish to show that R cannot be endowed with a compact topology. Suppose it could. Then R would be isomorphic to a power of F , say F^I , as F is semisimple, being a subdirect power of F , and therefore I. Kaplansky's structure theorem for compact semisimple rings [28, Theorem 16] applies (Theorem 8.21, derived independently of Kaplansky's theorem, also settles the matter). This would mean that R is isomorphic to the Boolean extension $F[P(I)]$, i.e.,

$$F[B] \cong F[P(I)] .$$

But now we can appeal to Theorem 3.5 of [9] which implies that two Boolean extensions of F can be isomorphic only if the Boolean algebras involved are themselves isomorphic. That B and $P(I)$ are not isomorphic is evident by the fact that $P(I)$ is atomic but B possesses no atoms whatsoever. Thus R is not compact.

Since the power set Boolean algebras are (up to isomorphism) precisely the complete atomic Boolean algebras, the argument given in Example 3 above actually shows that any ring R which is the Boolean extension of a finite field by a complete, non-atomic Boolean algebra is equationally compact but not compact. As a matter of fact, R is even a retract of a compact ring on account of a result of W. Taylor, which will be dealt with in the next section.

§8 Compactness

We turn now to discriminator varieties; in this paragraph we give characterizations of equational compactness and topological compactness in this setting. Briefly, the equationally compact members turn out to be products of Boolean extensions of finite simple members by complete Boolean algebras, and the compact members are the products of finite simple members.

8.1. Definition. The (ternary) discriminator on (the set) $\underline{A}$ is the ternary operation $t = t_A$ defined by

$$t(x,y,z) = \begin{cases} z & \text{if } x = y \\ x & \text{if } x \neq y \end{cases}, \quad x,y,z \in A .$$

The normal transform on $\underline{A}$ is the quaternary operation $n = n_A$ defined by

$$n(x,y,u,v) = \begin{cases} u & \text{if } x = y \\ v & \text{if } x \neq y \end{cases}, \quad x,y,u,v \in A .$$

A class $\mathbb{K}$ of algebras of type τ is said to be a discriminator class if there is a polynomial symbol $p \in P_3(\tau)$ inducing the discriminator on $\mathbb{K}$, that is satisfying $p^{\mathfrak{A}} = t_A$ for each $\mathfrak{A} \in \mathbb{K}$. If, moreoever, $\mathbb{K}$ is a finite set of finite algebras $\mathbb{K}$ is called quasi-primal. An algebra $\mathfrak{A}$ is a discriminator algebra resp. quasi-primal algebra if $\{\mathfrak{A}\}$ is a discriminator resp. quasi-primal class. Finally, an equational class $\mathbf{V}$ is a discriminator variety (resp. quasi-primal variety) if $\mathbf{V} = \mathbf{V}(\mathbb{K}) = \mathbf{HSP}(\mathbb{K})$ for some discriminator (resp. quasi-primal) class $\mathbb{K}$.

Note that the concepts defined in the second half of Definition 1 remain the same if everywhere the requirement of the "discriminator" is replaced by "normal transform"; this is evident from the relations $t(x,y,z) = n(x,y,z,x)$ and $n(x,y,u,v) = t(t(x,y,u),t(x,y,v),v)$. In the sequel we shall compute with t or with n as convenience demands.

In his original introduction and treatment of quasi-primal algebras A. Pixley's definition was different from, although equivalent to, our Definition 1. It soon became apparent that it was the ternary discriminator which was the key (and then not only for finite algebras alone!) to the far-reaching consequences for an equational class generated by a discriminator class. This is reflected in the following result of Pixley's which is most important for the investigations here, as it is what gives us access to the powerful "Lemma of Jónsson" (8.5).

<u>8.2. Proposition (A. Pixley [41, Lemma 2.3])</u>. An equational class $\mathbb{V}$ in $\mathbb{K}(\tau)$ is arithmetical (i.e., for every $\mathfrak{A} \in \mathbb{V}$ congruences on $\mathfrak{A}$ permute and $\mathcal{L}(\mathfrak{A})$ is a distributive lattice) if and only if there is a polynomial symbol $p \in P_3(\tau)$ such that $p(x,x,y) = p(y,x,x) = p(y,x,y) = y$ are identities of $\mathbb{V}$.

If $\mathbb{V} = \mathbb{V}(\mathbb{K})$ is a discriminator variety (whenever we write this in the future $\mathbb{K}$ will always be assumed to be a discriminator class generating $\mathbb{V}$ and t will denote the polynomial symbol inducing the discriminator on $\mathbb{K}$.) then every member of $\mathbb{K}$ satisfies the identities in the above with $p = t$, hence $\mathbb{V}$ satisfies the same identities; we have

8.3. Corollary. A discriminator variety is arithmetic.

A natural question arising at the outset is whether interesting examples of discriminator varieties exist; since our concern in this exposition is primarily ring theoretic we shall content ourselves with determining the discriminator varieties of rings, and refer the reader to [8] for a comprehensive list of examples including those of non-ring-type.

8.4. Proposition. For an equational class $\mathbb{V}$ of rings the following are equivalent:

(i) $\mathbb{V}$ is quasi-primal.

(ii) $\mathbb{V}$ is a discriminator variety.

(iii) $\mathbb{V} = \mathbb{V}(GF(q_1),\cdots,GF(q_n))$ for Galois fields $GF(q_1),\cdots, GF(q_n)$.

(iv) There is a natural number $m > 1$ such that $x^m = x$ is an identity of $\mathbb{V}$. (We call any ring satisfying this identity an m-ring.)

proof: (i) $\Rightarrow$ (ii) is trivial.
(ii) $\Rightarrow$ (iii): $\mathbb{V}$ is arithmetic by Corollary 3, and the arithmetic varieties of rings have been characterized by G. Michler and R. Wille [36, Satz 2] as being precisely those given by condition (iii).
(iii) $\Rightarrow$ (iv): The identity $x^{(q_1-1)(q_2-1)\cdots(q_n-1)+1} = x$ holds in each $GF(q_i)$, $i = 1,\cdots,n$, and hence in $\mathbb{V}$.
(iv) $\Rightarrow$ (iii): All members of $\mathbb{V}$ are commutative by a theorem of Jacobson, and all are semisimple, since obviously the von Neumann regularity condition (for every a there is an x fulfilling $axa = a$) holds. Hence any subdirectly irreducible

member of $\mathbf{V}$ is commutative and primitive, hence a field - moreover a field with not more than m elements, since it consists of roots of the equation $x^m - x = 0$. Thus $\mathbf{V}$ is generated by a finite number of Galois fields.

(iii) $\Rightarrow$ (1): The polynomial $z + (x - z)(x - y)^{(q_1-1)\cdots(q_n-1)}$ is the discriminator on each $GF(q_i)$, $i = 1,\cdots,n$. •

Note that if R is an m-ring then $\mathbf{V}(R)$ satisfies condition (iii), hence R is a member of a discriminator variety. In other words, <u>all results obtained for members of discriminator varieties are valid for all m-rings</u>.

Let us return now to the general setting and begin by quoting Jónsson's Lemma (mentioned earlier), which is valid in all discriminator varieties by virtue of Corollary 3.

<u>8.5. Proposition (B. Jónsson [27, Lemma 3.1 & Theorem 3.3])</u>.

(1) Let $\mathfrak{A}$ be a subalgebra of a direct product of algebras $\mathfrak{A}_i$, $i \in I$; suppose that $\mathcal{L}(\mathfrak{A})$ is distributive and that $\theta \in C(\mathfrak{A})$ is such that $\mathfrak{A}/\theta$ is subdirectly irreducible. Then there is an ultrafilter F on I satisfying $\theta_F|_A \leq \theta$.

(2) Let $\mathbb{V} = \mathbb{HSP}(\mathbb{K})$ be an equational class of algebras possessing distributive congruence lattices. Then

$$\mathbb{V} = \mathbb{IP}_s\mathbb{HSP}_u(\mathbb{K}) .$$

As a first application of the above we describe the subdirectly irreducibles in a discriminator variety.

<u>8.6. Proposition</u>. Let $\mathbf{V} = \mathbb{V}(\mathbb{K})$ be a discriminator variety with discriminator t. For $\mathfrak{A} \in \mathbf{V}$ with $|A| > 1$ the following

are equivalent:

(i) $\mathfrak{A}$ is subdirectly irreducible.

(ii) $\mathfrak{A} \in \mathbb{I}\mathbb{S}\mathbb{P}_u(\mathbb{K})$ ($\mathfrak{A} \in \mathbb{I}\mathbb{S}(\mathbb{K})$ if $\mathbb{V}$ is quasi-primal).

(iii) t induces the discriminator on $\mathfrak{A}$.

(iv) The discriminator operator t_A on A is a polynomial on $\mathfrak{A}$.

(v) $\mathfrak{A}$ is simple.

<u>proof</u>: (ii) ⇒ (iii): That the discriminator polynomial t induces, on a given member of $\mathbb{V}$, the discriminator operation, is describable by a first order sentence. Hence t is the discriminator on every ultraproduct of members of $\mathbb{K}$ by Łoś' Theorem (1.1), and therefore on any isomorphic copy of a subalgebra of same.

(iii) ⇒ (iv) is trivial.

(iv) ⇒ (v): Let $\Theta \in C(\mathfrak{A})$ and assume that $x \equiv y(\Theta)$ but $x \neq y$. Then for arbitrary $z \in A$, $z = t_A(x,x,z) \equiv t_A(x,y,z) = x$, hence $\Theta = 1$.

(v) ⇒ (i) is trivial.

(i) ⇒ (ii): By Proposition 5(2) $\mathbb{V} = \mathbb{I}\mathbb{P}_s\mathbb{H}\mathbb{S}\mathbb{P}_u(\mathbb{K})$, so assuming (i) $\mathfrak{A}$ must lie already in $\mathbb{H}\mathbb{S}\mathbb{P}_u(\mathbb{K})$. By the implication (ii) ⇒ (v) (already proved) $\mathbb{S}\mathbb{P}_u(\mathbb{K})$ consists of simple algebras, hence any homomorphic image of a member of this class with more than one element is a isomorphic copy. The statement in parentheses follows because $\mathbb{P}_u(\mathbb{K}) \subseteq \mathbb{I}(\mathbb{K})$ if $\mathbb{K}$ is quasi-primal. ●

Diverging somewhat from the approach given in [8] we fix the following terminology: From now on let $\mathbb{V} = \mathbb{V}(\mathbb{K})$ denote a discriminator variety; for $\mathfrak{A} \in \mathbb{V}$ let $P(\mathfrak{A})$ denote the set $\{\Theta(a,b);\ a,b \in A\}$ of principal congruences on $\mathfrak{A}$, and let

Spec $\mathfrak{A}$ be the set of (proper!) maximal congruences on $\mathfrak{A}$. It is immediate from Proposition 6 that every $\mathfrak{A} \in \mathbb{V}$ is a subdirect product of simple algebras, or, what is saying the same thing, every congruence on $\mathfrak{A}$ (in particular every principal congruence) is the meet of elements of Spec $\mathfrak{A}$. In the sequence of lemmata following we subject $\mathcal{L}(\mathfrak{A})$, $P(\mathfrak{A})$, and Spec $\mathfrak{A}$ to a closer scrutiny. The first result is really just the condensate of Jónsson's Lemma in discriminator varieties.

<u>8.7. Lemma</u>. Let $\mathfrak{A}_i$, $i \in I$, be simple algebras in $\mathbb{V}$ and let $\mathfrak{A} \subseteq \prod(\mathfrak{A}_i;\ i \in I)$. Then

$$\text{Spec}\,\mathfrak{A} = \{\theta_F|_A\ ;\ F \text{ is an ultrafilter on } I,\ \theta_F|_A \neq \iota\}.$$

<u>proof</u>: Let F be an ultrafilter on I. Then there is a natural embedding $\mathfrak{A}/\theta_F|_A \rightarrowtail \prod_F(\mathfrak{A}_i;\ i \in I)$. If $\theta_F|_A$ is not ι then $\mathfrak{A}/\theta_F|_A$ is simple by Proposition 6, and thus $\theta_F|_A \in$ Spec $\mathfrak{A}$. On the other hand, if $\theta \in$ Spec $\mathfrak{A}$, then by Proposition 5 (1) there exists an ultrafilter F on I with $\theta_F|_A \subseteq \theta$, hence $\theta_F|_A = \theta$ by maximality. ●

<u>8.8. Lemma</u>. For $\mathfrak{A} \in \mathbb{V}$ and $r,s,u,v \in A$

$$\theta(r,s) \leq \theta(u,v) \quad \text{iff} \quad t(u,v,r) = t(u,v,s).$$

Moreover, $P(\mathfrak{A})$ is a relatively complemented sublattice of $\mathcal{L}(\mathfrak{A})$, where the operations are computed by

$$\theta(r,s) \wedge \theta(u,v) = \theta(t(r,t(r,s,u),u),\ t(r,t(r,s,v),v),$$

$$\theta(r,s) \vee \theta(u,v) = \theta(t(r,s,u),\ t(s,r,v)),$$

and the relative complement of $\theta(r,s) \leq \theta(u,v)$ in the interval $[\omega, \theta(u,v)]$ is $\theta(t(r,s,u),\ t(r,s,v))$. It follows that

$\mathcal{P}(\mathfrak{A})$:= $\langle P(\mathfrak{A}); \cap, \vee \rangle$ is a Boolean sublattice of $\mathcal{L}(\mathfrak{A})$ iff ι is a principal congruence.

proof: For $\phi \in \operatorname{Spec}\mathfrak{A}$ and $x \in A$ abbreviate $[x]\phi$ by $\bar{x}$. Keeping Proposition 6 in mind every step of the way, note first that

$$(*) \qquad \Theta(x,y) = \bigcap(\phi;\ \phi \in \operatorname{Spec}\mathfrak{A} \text{ and } \bar{x} = \bar{y}) .$$

It follows that $\Theta(r,s) \leq \Theta(u,v)$ iff

$$\bar{u} = \bar{v} \Rightarrow \bar{r} = \bar{s}, \quad \text{all } \phi \in \operatorname{Spec}\mathfrak{A}$$

$$\text{iff} \quad t(\bar{u},\bar{v},\bar{r}) = t(\bar{u},\bar{v},\bar{s}), \quad \text{all } \phi \in \operatorname{Spec}\mathfrak{A}$$

$$\text{iff} \quad \overline{t(u,v,r)} = \overline{t(u,v,s)}, \quad \text{all } \phi \in \operatorname{Spec}\mathfrak{A}$$

iff $t(u,v,r) = t(u,v,s)$, which is the first statement. Now by $(*)$ again,

$$\Theta(r,s) \cap \Theta(u,v) = \bigcap(\phi \in \operatorname{Spec}\mathfrak{A};\ \bar{r}=\bar{s}) \cap \bigcap(\phi \in \operatorname{Spec}\mathfrak{A};\ \bar{u}=\bar{v})$$

$$= \bigcap(\phi \in \operatorname{Spec}\mathfrak{A};\ \bar{r} = \bar{s} \text{ or } \bar{u} = \bar{v}) .$$

But for $\phi \in \operatorname{Spec}\mathfrak{A}$, $\bar{r} = \bar{s}$ or $\bar{u} = \bar{v}$ iff $t(\bar{r},t(\bar{r},\bar{s},\bar{u}),\bar{u}) = t(\bar{r},t(\bar{r},\bar{s},\bar{v}),\bar{v})$, as is easily checked. Hence

$$\Theta(r,s) \cap \Theta(u,v) = \bigcap(\phi \in \operatorname{Spec}\mathfrak{A};\ \overline{t(r,t(r,s,u),u)} = \overline{t(r,t(r,s,v),v)})$$

$$= \Theta(t(r,t(r,s,u),u),\ t(r,t(r,s,v),v))$$

by $(*)$. Similarly, note that

$$\Theta(r,s) \vee \Theta(u,v) = \bigcap(\phi \in \operatorname{Spec}\mathfrak{A};\ \bar{r} = \bar{s} \text{ and } \bar{u} = \bar{v})$$

and check that $\bar{r} = \bar{s}$ and $\bar{u} = \bar{v}$ iff $t(\bar{r},\bar{s},\bar{u}) = t(\bar{s},\bar{r},\bar{v})$, verifying the join relation. Using the relations won above we compute (under the assumption that $\Theta(r,s) \leq \Theta(u,v)$):

$$\Theta(r,s) \vee \Theta(t(r,s,u),t(r,s,v))$$

$$= \Theta(t(r,s,t(r,s,u)),\ t(s,r,t(r,s,v)))$$

$$= \Theta(t(r,s,u),\ t(s,r,v))$$

(since for all $\phi \in \operatorname{Spec}\mathfrak{A}$, $t(\bar{r},\bar{s},t(\bar{r},\bar{s},\bar{u})) = t(\bar{r},\bar{s},\bar{u})$ and

$t(\bar{s},\bar{r},t(\bar{r},\bar{s},\bar{v})) = t(\bar{s},\bar{r},\bar{v})$)

$$= \Theta(r,s) \vee \Theta(u,v) = \Theta(u,v) \; ; \quad \text{and}$$

$$\Theta(r,s) \wedge \Theta(t(r,s,u),\ t(r,s,v))$$

$$= \Theta(t(r,t(r,s,t(r,s,u)),t(r,s,u)),\ t(r,t(r,s,t(r,s,v),t(r,s,v)))$$

$$= \Theta(r,r) = id_A = \omega . \quad \bullet$$

8.9. Lemma. For $\mathfrak{A} \in \mathbf{V}$ each $\Theta(u,v) \in P(\mathfrak{A})$ is closed, and if $\mathfrak{A}$ is equationally compact, $\mathfrak{P}(\mathfrak{A})$ is a complete Boolean lattice and is a sublattice of $\mathfrak{L}(\mathfrak{A})$ (i.e., $\iota \in P(\mathfrak{A})$).

proof: By Lemma 8 $r \equiv s(\Theta(u,v))$ iff $t(u,v,r) = t(u,v,s)$, which implies that $\Theta(u,v)$ is the solution set in A^2 of the equation $t(u,v,x_o) = t(u,v,x_1)$; thus $\Theta(u,v)$ is closed. Now suppose $\mathfrak{A}$ is equationally compact. Let $X = \{\Theta(r_i,s_i);\ i \in I\}$ be an arbitrary subset of $P(\mathfrak{A})$. We wish to show that sup X and inf X exist in $\mathfrak{P}(\mathfrak{A})$. If $\{\Theta(u_j,v_j);\ j \in J\}$ is the set of upper bounds of X in $\mathfrak{P}(\mathfrak{A})$, let

$$\Sigma := \{\Theta(r_i,s_i) \leq \Theta(x_o,x_1);\ i \in I\} \cup \{\Theta(x_o,x_1) \leq \Theta(u_j,v_j);\ j \in J\} .$$

Then sup X exists if there exist $x_o, x_1 \in A$ fulfilling Σ , and this holds if there is a solution in $\mathfrak{A}$ (for x_o and x_1) of the system of polynomial equations

$$\Sigma_{\mathfrak{A}} := \{t(x_o,x_1,r_i) = t(x_o,x_1,s_i) \; ; \; i \in I\}$$
$$\cup \{t(u_j,v_j,x_o) = t(u_j,v_j x_1) \; ; \; j \in J\} ,$$

because of Lemma 8 again. As $\mathfrak{P}(\mathfrak{A})$ is a lattice any finite set of the conditions Σ can be fulfilled simultaneously in $\mathfrak{P}(\mathfrak{A})$, hence the corresponding subset of $\Sigma_{\mathfrak{A}}$ is solvable in $\mathfrak{A}$. Since $\mathfrak{A}$ is equationally compact $\Sigma_{\mathfrak{A}}$ is solvable and therefore sup X exists. The argument for inf X is analogous. Now since <u>any</u> congruence is the (infinite) join in $\mathcal{L}(\mathfrak{A})$ of the principal congruences lying under it, we conclude, in particular, that ι is principal. ●

A word of caution regarding the above: if $\mathfrak{A}$ is equationally compact, we have $\mathcal{P}(\mathfrak{A})$ a 0,1 - sublattice of $\mathcal{L}(\mathfrak{A})$ and both are complete; but infinite joins do <u>not</u>, in general, agree - for example, the ring R constructed in Example 7.3 is an equationally compact m-ring which is not compact. If infinite joins in $\mathcal{P}(R)$ coincided with those in $\mathcal{L}(R)$, then R would be a principal ideal ring, therefore compact after all by Theorem 6.5, a contradiction. Where confusion may arise we will write $\bigvee_C$ or $\bigvee_P$ to indicate in which lattice the sup is taken.

For $r,s \in A$ we abbreviate $\{\phi \in \operatorname{Spec}\mathfrak{A} \,;\, r \equiv s(\phi)\}$ by <u>E(r,s)</u> and $\operatorname{Spec}\mathfrak{A} \setminus E(r,s)$ by <u>D(r,s)</u> .

<u>8.10. Lemma</u>. Let $\mathfrak{A} \in \mathbf{V}$. Define :

$$j : C(\mathfrak{A}) \to 2^{\operatorname{Spec}\mathfrak{A}}$$
$$\theta \mapsto \{\phi \in \operatorname{Spec}\mathfrak{A} \,;\, \phi \geq \theta\} .$$

Then

(1) $j(\omega) = \operatorname{Spec}\mathfrak{A}$ and $j(\iota) = \emptyset$.

(2) j is injective.

(3) $j(\theta(r,s) \wedge \theta(u,v)) = j(\theta(r,s)) \cup j(\theta(u,v))$.

(4) $j(\bigvee_C(\theta(r_i,s_i);\, i \in I)) = \bigwedge(j(\theta(r_i,s_i));\, i \in I)$.

(5) If $\iota \in P(\mathfrak{A})$ then

$$j : \mathcal{P}(\mathfrak{A}) \to 2^{\operatorname{Spec}\mathfrak{A}}$$

is an anti-monomorphism of Boolean algebras.

<u>proof</u>: (1) is obvious.

(2): As every congruence is the meet of all the maximal ones larger than it, it follows from $j(\Theta) = j(\phi)$ that

$$\Theta = \bigcap j(\Theta) = \bigcap j(\phi) = \phi .$$

(3): In the proof of Lemma 8 we saw that for $\phi \in \operatorname{Spec} \mathfrak{A}$ $\bar{r} = \bar{s}$ or $\bar{u} = \bar{v}$ iff $\overline{t(r,t(r,s,u),u)} = \overline{t(r,t(r,s,v),v)}$ (and hence $\Theta(r,s) \cap \Theta(u,v) = \Theta(t(r,t(r,s,u),u),\ t(r,t(r,s,v),v))$). Therefore $\phi \in j(\Theta(r,s) \cap \Theta(u,v))$ iff $\phi \geq \Theta(t(r,t(r,s,u),u),$ $t(r,t(r,s,v),v))$ iff $\phi \geq \Theta(r,s)$ or $\phi \geq \Theta(u,v)$ iff $\phi \in j(\Theta(r,s)) \cup j(\Theta(u,v))$.

(4): $\phi \in j(\bigvee(\Theta(r_i,s_i);\ i \in I))$ iff $\phi \in \bigvee(\Theta(r_i,s_i);\ i \in I)$ iff for all $i \in I$ $\phi \geq \Theta(r_i,s_i)$ iff $\phi \in \bigcap(j(\Theta(r_i,s_i));\ i \in I)$.

(5) follows directly from (1) - (4) and Lemma 8. ●

8.11. Proposition. Let $\mathfrak{A} \in \mathbb{V}$ with $1 \in P(\mathfrak{A})$. Then with

$$j(P(\mathfrak{A})) = \{E(r,s);\ r,s \in A\}$$

as a basis for the open sets, $\operatorname{Spec} \mathfrak{A}$ becomes a Boolean space. Moreover, $j(P(\mathfrak{A}))$ is then the Boolean algebra of clopen sets of $\operatorname{Spec} \mathfrak{A}$, and the map

$$\begin{aligned} j' : C(\mathfrak{A}) &\rightarrow 2^{\operatorname{Spec} \mathfrak{A}} \\ \Theta &\mapsto (\operatorname{Spec} \mathfrak{A}) \setminus j(\Theta) \end{aligned}$$

yields a lattice isomorphism between $\mathcal{L}(\mathfrak{A})$ and the lattice of open sets of $\operatorname{Spec} \mathfrak{A}$.

proof: For $\Theta, \phi \in \operatorname{Spec} \mathfrak{A}$ with $\Theta \neq \phi$ there exist (w.l.o.g.) $r,s \in A$ satisfying $r \equiv s(\Theta)$ but $r \not\equiv s(\phi)$. As $j(P(\mathfrak{A}))$ is a sub-Boolean-algebra of $2^{\operatorname{Spec} \mathfrak{A}}$ (Lemma 10 (5)), $E(r,s)$ and $D(r,s)$ are two open sets separating Θ and ϕ ; thus $\operatorname{Spec} \mathfrak{A}$ is Hausdorff. Total disconnectedness of $\operatorname{Spec} \mathfrak{A}$ follows immediately from Hausdorffness and the fact that the basic open sets are clopen. Now the family $j(P(\mathfrak{A}))$ is also a basis of closed sets for Spec $\mathfrak{A}$, hence, to verify that Spec $\mathfrak{A}$ is compact, we need to show that under the assumption

$$\bigcap(j(\Theta(r_i,s_i));\ i\in I) = \emptyset ,$$

some finite intersection of the above family must be empty. Appealing to Lemma 10 we have

$$j(\nabla) = \emptyset = \bigcap(j(\Theta(r_i,s_i));\ i\in I) = j(\bigvee_C(\Theta(r_i,s_i);\ i\in I))$$

hence

$$\nabla = \bigvee_C(\Theta(r_i,s_i);\ i\in I) .$$

Since the principal congruences are compact elements of the algebraic lattice $\mathcal{L}(\mathfrak{A})$ and $\nabla\in P(\mathfrak{A})$ it follows that

$$\nabla = \Theta(r_{i_1},s_{i_1})\vee\cdots\vee\Theta(r_{i_m},s_{i_m})$$

for suitable $i_1,\cdots,i_m\in I$, and hence

$$\emptyset = j(\nabla) = j(\Theta(r_{i_1},s_{i_1})\vee\cdots\vee\Theta(r_{i_m},s_{i_m}))$$

$$= j(\Theta(r_{i_1},s_{i_1}))\cap\cdots\cap j(\Theta(r_{i_m},s_{i_m}))$$

as desired. Thus Spec $\mathfrak{A}$ is compact, and thus a Boolean space. Now by construction every $E(r,s)$ is clopen. Conversely, suppose $M\subseteq$ Spec $\mathfrak{A}$ is clopen. Then M is compact, being closed, and

$$M = \bigcup(j(\Theta(r_i,s_i));\ i\in I)$$

being open. Then by compactness and Lemma 10 there are $i_1,\cdots,i_m\in I$ with

$$M = \bigcup(j(\Theta(r_{i_k},s_{i_k}));\ k=1,\cdots,m)$$

$$= j(\bigcap(\Theta(r_{i_k},s_{i_k});\ k=1,\cdots,m) ,$$

hence $M\in j(P(\mathfrak{A}))$.

Now j' is order-preserving as j is anti-order-preserving (Lemma 10) and, moreover, an injection. As $\bigcap(j(\theta(r_i,s_i)); i \in I)$ is a typical closed set in $\mathrm{Spec}\,\mathfrak{A}$ and $\bigvee_C(\theta(r_i,s_i); i \in I)$ is a typical congruence on $\mathfrak{A}$, it follows by Lemma 10 (4) that j' is a bijection between $\mathcal{L}(\mathfrak{A})$ and the lattice of open sets of $\mathrm{Spec}\,\mathfrak{A}$. But an order isomorphism between two lattices is a lattice isomorphism. •

The next result is, in essence, an important theorem of W. Taylor [49, Theorem 1.6], which will give us, among other things, the key to the answer to the Mycielski question in discriminator varieties. The form of the proof we give is taken from [8], although the formulation of Taylor's theorem given there appears to be over-stated.

<u>8.12. Proposition</u>. For $\mathfrak{A} \in \mathbb{V}$ the canonical embedding

$$f : \mathfrak{A} \rightarrowtail \prod(\mathfrak{A}/\theta \;;\; \theta \in \mathrm{Spec}\,\mathfrak{A})$$

is pure.

<u>proof</u>: If $p = q$ is a polynomial equation in the variables $x_1,\cdots,x_n$, then one easily checks (using the accustomed trick of checking on the simple quotients) that $p = q$ is equivalent to the equation $n(p,q,x_1,t(p,x_1,q)) = x_1$; moreover, the conjunction of two polynomial equations of the form $p_1 = x_1$ and $p_2 = x_1$ in the variables $x_1,\cdots,x_n$ is equivalent to the polynomial equation $t(p_1,x_1,p_2) = x_1$. Therefore by induction the conjunction of a finite number of polynomial equations over $\mathfrak{A}$ in the variables $x_1,\cdots,x_n$ is equivalent to an equation of the form

$$p(x_1,\cdots,x_n) = x_1$$

with constants in $\mathfrak{A}$. To verify that f is pure we must consider an equation

(1) $$p(x_1,\cdots,x_m,a_1,\cdots,a_k) = x_1$$

where p is a polynomial in $m+k$ variables and $a_1,\cdots,a_k \in A$, and show, under the assumption that (1) is not solvable in $\mathfrak{A}$, that then

(2) $$p(x_1,\cdots,x_m,f(a_1),\cdots,f(a_k)) = x_1$$

is not solvable in $\mathfrak{B} := \prod(\mathfrak{A}/\theta\ ;\ \theta \in \operatorname{Spec}\mathfrak{A})$. Write for brevity $(u) = (u_1,\cdots,u_l)$ for any variable tuples or tuples of constants and correspondingly $[(u)]\theta$ for the tuple $([u_1]\theta,\cdots,[u_l]\theta)$. For a <u>pregiven</u> constant r_j let (r_j) denote the l-tuple $(r_j,\cdots,r_j)$. (The value of l will always be clear from the context.) Define, for all $(s) \in A^m$,

$$L_{(s)} := \{\theta \in \operatorname{Spec}\mathfrak{A}\ ;\ p([(s)]\theta,[(a)]\theta) = [s_1]\theta\}.$$

An easy computation shows that

$$L_{(s)} \cup L_{(t)} = L_{n((p((s),(a))),(s_1),(s),(t))}$$

from which it follows that

$$I := \{M \subseteq \operatorname{Spec}\mathfrak{A}\ ;\ M \subseteq L_{(s)} \text{ for some } (s) \in A^m\}$$

is an ideal on $\operatorname{Spec}\mathfrak{A}$. Under the assumption that (1) is not solvable in $\mathfrak{A}$ it follows that $\operatorname{Spec}\mathfrak{A} \notin I$, i.e., I is a proper ideal, and therefore there exists an ultrafilter F on $\operatorname{Spec}\mathfrak{A}$ with $I \cap F = \emptyset$. By Lemma 7 and the fact that $f(\mathfrak{A})$ is subdirect in $\mathfrak{B}$ we know that $\theta_F|_{f(A)} \in \operatorname{Spec} f(\mathfrak{A})$; let $\theta(F) \in \operatorname{Spec}\mathfrak{A}$ denote the kernel of the composition

$$\mathfrak{A} \overset{f}{\rightarrowtail\!\!\!\twoheadrightarrow} f(\mathfrak{A}) \twoheadrightarrow f(\mathfrak{A})/\theta_F|_{f(A)}.$$

Suppose now that (2) has a solution in $\mathfrak{B}$, say $(s) = (s_1,\cdots,s_m) \in B^m$. Choose $(t) = (t_1,\cdots,t_m) \in A^m$ satisfying, for each $i = 1,\cdots,m$, $[t_i]\theta(F) = s_i(\theta(F))$. As (s) was a solution, it is a solution on every component, which implies in particular that

$$p((t),(a)) \equiv t_1 \pmod{\theta(F)} ,$$

which, by definition of a filter congruence, yields that F must contain the set

$$\{\theta \in \operatorname{Spec} \mathfrak{A} \ ; \ p([(t)]\theta, \ [(a)]\theta) = [t_1]\theta$$

which is just $L_{(t)}$, contradicting the choice of F . ●

In order to proceed we must demand that "decency" condition on our discriminator variety $\mathbb{V}$ hinted at, but not formulated, in [8] as well as in our introductory comments at the beginning of the chapter. It is the following, very mild restriction, which we will tacitly assume to hold on $\mathbb{V}$ for the rest of the chapter :

(D) For every $n \in \mathbb{N}$ there are, up to isomorphism, only a finite number of simple algebras in $\mathbb{V}$ with n elements.

Of course (D) asserts no restrictions on quasi-primal varieties nor on (discriminator) varieties of a finite type τ . To emphasize its validity at times we will refer to $\mathbb{V}$ as being a "(D)-discriminator) variety". Its introduction at this point is prompted by the fact that the preceding general investigation has not depended on (D) , but the following Proposition 12, crucial for the main results, is definitely false without it.

For $\mathfrak{A} \in \mathbb{V}$ and any simple algebra $\gamma \in \mathbb{V}$ denote by $M_\gamma(\mathfrak{A})$ the set $\{\phi \in \operatorname{Spec} \mathfrak{A} \ ; \ \mathfrak{A}/\phi \cong \gamma\}$,

<u>8.13. Proposition</u>. Let $\{\mathfrak{A}_i ; i \in I\}$ be a set of pairwise non-isomorphic, non-trivial simple algebras in $\mathbb{V}$. For $i \in I$ let K_i be a set, let $K := \dot{\bigcup}(K_i ; i \in I)$, and set

$$\mathfrak{A} := \prod(\mathfrak{A}_i^{K_i} ; i \in I) .$$

Then $M_{\mathfrak{A}_i}(\mathfrak{A})$ is clopen for all finite $\mathfrak{A}_i$.

<u>proof</u>: Note that the topological statement is meaningful, as $1 \in P(\mathfrak{A})$ by Lemma 9. By Lemma 7 we have

$$M_{\mathfrak{A}_i}(\mathfrak{A}) = \{\theta \in \operatorname{Spec}\mathfrak{A} ; \mathfrak{A}/\theta \cong \mathfrak{A}_i\}$$

$$= \{\theta_F ; F \text{ is an ultrafilter on } K \text{ with } \mathfrak{A}/\theta_F \cong \mathfrak{A}_i\} .$$

Let $\mathfrak{A}_i$ be finite. We claim that, for an ultrafilter F on K, $\mathfrak{A}/\theta_F \cong \mathfrak{A}_i$ iff $K_i \in F$. Indeed, if $K_i \in F$ then $\mathfrak{A}/\theta_F = \prod_F(\mathfrak{A}_i^{K_i} ; i \in I) \cong (\mathfrak{A}_i)_{F_i}^{K_i} \cong \mathfrak{A}_i$, where F_i is the restriction of F to K_i. Conversely, assume that $\mathfrak{A}/\theta_F \cong \mathfrak{A}_i$. Then in particular $\mathfrak{A}/\theta_F$ has cardinality $n = |A_i|$. Having the finite cardinality n is a first order property, so by Łoś' Theorem (1.1)

$$\dot{\bigcup}(K_j ; |A_j| = n) \in F .$$

But condition (D) insures that only a finite number of K_j's appear in the above union, hence for some $j \in I$, $K_j \in F$. Thus $\mathfrak{A}/\theta_F \cong \mathfrak{A}_j$ by our first argument, whence $j = i$, as the $\mathfrak{A}_j$'s were chosen pairwise non-isomorphic, and so $K_i \in F$ as claimed. We conclude that

$$M_{\mathfrak{A}_i}(\mathfrak{A}) = \{\theta_F ; F \text{ is an ultrafilter on } K \text{ with } K_i \in F\}.$$

Fix $r,s \in A$ satisfying $\{k \in K; r(k) = s(k)\} = K_i$. Then for any ultrafilter F on K, $r \equiv s(\theta_F)$ iff $K_i \in F$, and therefore

$$M_{\mathfrak{a}_i}(\mathfrak{a}) = \{\theta_F \; ; \; F = \text{ultrafilter on } K \text{ with } r \equiv s(\theta_F)\}$$

$$= \{\phi \in \operatorname{Spec} \mathfrak{a} \; ; \; r \equiv s(\phi)\} = E(r,s) ,$$

by appealing again to Lemma 7. Thus $M_{\mathfrak{a}_i}(\mathfrak{a})$ is clopen. •

8.14. Lemma. Let $\mathfrak{a} \in \mathbb{V}$, $\tau \in P(\mathfrak{a})$, and let $\hat{\mathfrak{a}}$ be the image of $\mathfrak{a}$ under the canonical embedding

$$\mathfrak{a} \rightarrowtail \prod(\mathfrak{a}/\theta; \; \theta \in \operatorname{Spec} \mathfrak{a})$$

$$a \longmapsto \hat{a} = ([a]\theta)_{\theta \in \operatorname{Spec} \mathfrak{a}} .$$

Then given any pairwise disjoint clopen sets $O_1, \cdots, O_n$ in $\operatorname{Spec} \mathfrak{a}$, and any $\hat{r}_1, \cdots, \hat{r}_n \in \hat{A}$, there exists $\hat{r} \in \hat{A}$ such that

$$\hat{r}|_{O_i} = \hat{r}_i|_{O_i} , \quad i = 1, \cdots, n .$$

proof: By induction it suffices to verify the claim for $n = 2$. As is easily checked, taking $\hat{r}$ given by $r = n(a,b,r_1,r_2)$, where $O_1 = E(a,b)$, does the trick. •

We will refer to an r satisfying the above as a "gluing together of $r_1, \cdots, r_n$ over $O_1, \cdots, O_n$".

8.15. Lemma. Let $\mathfrak{a} \in \mathbb{V}$ with $\tau \in P(\mathfrak{a})$. Let $n \in \mathbb{N}$ and suppose $E(r,s)$ is a clopen set in $\operatorname{Spec} \mathfrak{a}$ such that for all $\theta \in E(r,s)$ $|A/\theta| \geq n$. Then there exist $a_1, \cdots, a_n \in A$ which are "disjoint over $E(r,s)$", meaning

$$E(a_i,a_j) \cap E(r,s) = \emptyset \quad \text{whenever } i \neq j .$$

proof: By hypothesis we can choose, for each $\phi \in E(r,s)$, elements $a_1^{(\phi)}, \cdots, a_n^{(\phi)}$ in A satisfying $a_i^{(\phi)} \not\equiv a_j^{(\phi)} \; (\phi)$ for $i \neq j$. Then $O_\phi := \bigcap(D(a_i,a_j); \; i \neq j)$ is a clopen set containing ϕ , hence

$$E(r,s) \subseteq \bigcup(O_\phi \ ; \ \phi \in E(r,s)) \ .$$

As $E(r,s)$ is compact, there is a finite subcovering

$$E(r,s) \subseteq O_{\phi_1} \cup \cdots \cup O_{\phi_m} \ .$$

All sets in the above are clopen, so this covering can be disjointified, that is, there are clopen sets $O_1 \subseteq O_{\phi_1}$, $\cdots$, $O_m \subseteq O_{\phi_m}$ with $E(r,s) = O_1 \dot\cup \cdots \dot\cup O_m$. For each $i=1,\cdots,n$ let a_i be obtained by gluing together $a_i^{(\phi_1)},\cdots, a_i^{(\phi_m)}$ over $O_1,\cdots, O_m$. It follows from the definition of the O_ϕ's that $a_1,\cdots, a_n$ have the desired property. ●

We now have sufficient tools at our disposal to attack the first characterization theorem.

<u>8.16. Theorem</u>. Let $\mathbb{V}$ be a (D)-discriminator variety and let $\mathfrak{A} \in \mathbb{V}$. Then $\mathfrak{A}$ is equationally compact iff there are pairwise non-isomorphic finite simple algebras $\mathfrak{A}_i \in \mathbb{V}$, $i \in I$, and complete Boolean algebras B_i , $i \in I$, such that

$$\mathfrak{A} \ \cong \ \prod(\mathfrak{A}_i[B_i] \ ; \ i \in I) \ .$$

<u>proof</u>: The sufficiency of the decomposition property follows immediately from Theorem 7.2. So assume that $\mathfrak{A}$ is equationally compact. By Lemma 9 $\imath \in P(\mathfrak{A})$, so all the results depending on this condition are valid for $\mathfrak{A}$. We have the canonical embedding

$$\mathfrak{A} \ \rightarrowtail \ \prod(\mathfrak{A}/\theta \ ; \ \theta \in \operatorname{Spec}\mathfrak{A}\ ) \ =: \ \mathfrak{B}' \ ,$$

which is pure by Proposition 12. Hence by Weglorz' Theorem (2.3) there exists a retraction

$$g' : \mathfrak{B}' \ \rightarrow \ \mathfrak{A} .$$

Now $\mathcal{B}'$ is isomorphic to a product

$$\prod(\mathfrak{A}_j^{K_j};\ j \in J) \quad =: \mathcal{B} ,$$

where $\mathfrak{A}_j$, $j \in J$, are pairwise non-isomorphic non-trivial simple algebras; hence there is an epimorphism

$$g : \mathcal{B} \twoheadrightarrow \mathfrak{A} .$$

Let I be that subset of J indexing the <u>finite</u> members $\mathfrak{A}_j$; by Proposition 13 $M_{\mathfrak{A}_i}(\mathcal{B})$ is clopen for $i \in I$, say $M_{\mathfrak{A}_i}(\mathcal{B}) = E(a_i, b_i)$, $i \in I$. We claim that $M_{\mathfrak{A}_i}(\mathfrak{A})$ is clopen too, namely, $M_{\mathfrak{A}_i}(\mathfrak{A}) = E(g(a_i), g(b_i))$. Indeed, suppose $\phi \in \operatorname{Spec} \mathfrak{A}$. If p denotes the canonical epimorphism of $\mathfrak{A}$ onto $\mathfrak{A}/\phi$, then $\mathfrak{A}/\phi \cong \mathfrak{A}_i$ iff $\mathcal{B}/\ker(p \circ g) = \mathfrak{A}_i$ iff $a_i \equiv b_i(\ker(p\circ g))$ iff $p(g(a_i)) = p(g(b_i))$ iff $g(a_i) \equiv g(b_i)(\phi)$. Setting $r_i = g(a_i)$ and $s_i = g(b_i)$, $i \in I$, we may sum up :

$$M_i \ := \ M_{\mathfrak{A}_i}(\mathfrak{A}) \ = \ E(r_i, s_i) , \quad i \in I .$$

Using the terminology of Lemma 14 let $\mathfrak{S}_i$ be the image of the canonical projection p_i of $\hat{\mathfrak{A}}$ onto those components indexed by the set M_i , $i \in I$, and denote by $\hat{a}^i$ the element $p_i(a) = ([a]\theta)_{\theta \in M_i}$ for $a \in A$. Note that $\mathfrak{S}_i \cong \mathfrak{A}/\theta(r_i, s_i)$. Consider now the homomorphism

$$f : \mathfrak{A} \longrightarrow \prod(\mathfrak{S}_i \ ; \ i \in I) .$$
$$a \longmapsto (\hat{a}^i)_{i \in I}$$

We shall show that f is an isomorphism. Now an arbitrary element of $\prod(S_i ;\ i \in I)$ is of the form $(\hat{a}_i^i)_{i \in I}$ with $a_i \in A$. To show that f is surjective we must find an element $a \in A$ satisfying $\hat{a}^i = \hat{a}_i^i$ for all $i \in I$, i.e., satisfying $a \equiv a_i(\phi)$ for $i \in I$ and $\phi \in M_i$, i.e., such that the implication

$$r_i \equiv s_i \ (\phi) \Rightarrow a \equiv a_i \ (\phi) , \quad i \in I , \quad \phi \in \operatorname{Spec} \mathfrak{A}$$

holds. To this end consider the following system of equations with constants in $\mathfrak{A}$:

$$\Sigma = \{t(r_i,s_i,x) = t(r_i,s_i,a_i) \; ; \; i \in I\} .$$

Σ is finitely solvable in $\mathfrak{A}$. Indeed, if $i_1,\cdots,i_m$ are the indices occurring in a finite subset Σ_o of Σ, then a solution for Σ_o is obtained by gluing together $a_{i_1},\cdots,a_{i_m}$ over $E(r_{i_1},s_{i_1}), \cdots, E(r_{i_m},s_{i_m})$ (Lemma 14). As $\mathfrak{A}$ is equationally compact there is a solution $a \in A$ for Σ, and clearly $f(a) = (\hat{a}_i^i)_{i\in I}$. Thus f is surjective.

Now let $a,b \in A$ and assume that $f(a) = f(b)$. To show is that $a = b$. Consider the system of equations

$$\Sigma = \{t(x_p,x_q,a) = x_p = t(x_p,x_q,b) \; ; \; p,q \in P \; , \; p \neq q\} ,$$

where the index set P is chosen such that $|P| > |A|$. Now Σ is finitely solvable in $\mathfrak{A}$. Indeed, let $p_1,\cdots,p_m$ be the indices occurring in a finite subset Σ_o of Σ; since $f(a) = f(b)$ it follows that for any $\phi \in D(a,b)$ the quotient $\mathfrak{A}/\phi$ is infinite, and therefore by Lemma 15 there are elements $a_1',\cdots,a_m'$ which are disjoint over the clopen set $D(a,b)$. If, for $i = 1,\cdots,m$, a_i is obtained by gluing together a, a_i' over $E(a,b)$, $D(a,b)$, then the substitution $x_{p_i} \mapsto a_i$ is a solution of Σ_o. Hence there is a solution $(c_p)_{p\in P}$ of Σ in $\mathfrak{A}$. As P was chosen larger than A, $c_p = c_q$ for some $p \neq q$. But then

$$a = t(c_p,c_q,a) = c_p = t(c_p,c_q,b) = b ,$$

as $t(u,u,v) = v$ is an identity on $\mathfrak{A}$. Thus $a = b$ and f is injective as claimed. We give at this point a formulation of what we are in the process of proving in which the technical details of the proof are set off in such a way as to save having to refer to the proof per se at times later on in the discussion.

8.17. Lemma. If $\mathfrak{A} \in \mathbb{V}$ is equationally compact, then $\mathfrak{A} \cong \prod(\mathfrak{T}_i;\ i \in I)$, where $\mathfrak{T}_i \cong \mathfrak{A}/\theta(r_i,s_i)$ and where, moreover, there are pairwise non-isomorphic finite simple algebras $\mathfrak{A}_i$ such that $M_{\mathfrak{A}_i}(\mathfrak{A}) = E(r_i,s_i) =: M_i$ and $\mathfrak{T}_i \cong \mathfrak{A}_i[B_i]$ with B_i the complete Boolean algebra of clopen sets in $\operatorname{Spec}\mathfrak{T}_i$.

By verifying the very last statement of the above, its proof, as well as that of the theorem, will be completed. Now the congruence $\theta(r_i,s_i)$ is closed by Lemma 9 and therefore $\mathfrak{T}_i \cong \mathfrak{A}/\theta(r_i,s_i)$ is equationally compact by Proposition 2.8. Note also that if $\phi \in \operatorname{Spec}(\mathfrak{A}/\theta(r_i,s_i))$ then $\phi = \phi'/\theta(r_i,s_i)$ with $\phi' \geq \theta(r_i,s_i)$, i.e., with $\phi' \in E(r_i,s_i) = M_{\mathfrak{A}_i}(\mathfrak{A})$, and hence

$$\mathfrak{A}/\theta(r_i,s_i)/\phi \;\cong\; \mathfrak{A}/\theta(r_i,s_i)/\phi'/\theta(r_i,s_i) \;\cong\; \mathfrak{A}/\phi' \;\cong\; \mathfrak{A}_i\ .$$

We conclude that every proper simple quotient of $\mathfrak{T}_i$ is isomorphic to $\mathfrak{A}_i$. In other words, we may assume w.l.o.g. that $\mathfrak{A}$ is an equationally compact algebra satisfying $\mathfrak{A}/\phi \cong \mathfrak{A}_0$ for every $\phi \in \operatorname{Spec}\mathfrak{A}$, where $\mathfrak{A}_0$ is a fixed finite simple algebra, and we must show that $\mathfrak{A} \cong \mathfrak{A}_0[B]$, where B is the complete (!) Boolean algebra of clopen sets in $\operatorname{Spec}\mathfrak{A}$.

First off, B <u>is</u> complete by Lemma 9, Lemma 10 (5) and Proposition 11. Fixing $\theta_0 \in \operatorname{Spec}\mathfrak{A}$ we may assume w.l.o.g. that $\mathfrak{A}_0$ is $\mathfrak{A}/\theta_0$. Say $|A_0| = n$. Then by Lemma 15 there are $a_1, \cdots, a_n \in A$ disjoint over $\operatorname{Spec}\mathfrak{A}$, that is, satisfying

$$E(a_i,a_j) = \emptyset \quad \text{for } i \neq j\ .$$

Define

$$h : \mathfrak{A} \longrightarrow \mathfrak{A}_o^{\operatorname{Spec}\mathfrak{A}}$$
$$a \longmapsto \hat{\hat{a}}$$

where $\hat{\hat{a}}(\phi) = [a_i]\Theta_o$ iff $a \equiv a_i(\phi)$. h is obviously a monomorphism; we will be finished if we show that $h(A)$ is the set of continuous maps from $\operatorname{Spec}\mathfrak{A}$ to $\mathfrak{A}_o$ (since by Proposition 11 $\operatorname{Spec}\mathfrak{A}$ is the Stone space of B). Now for $a \in A$, $\hat{\hat{a}}^{-1}([a_i]\Theta_o) = E(a,a_i)$, hence each $\hat{\hat{a}}$ is continuous. Conversely, suppose $s \in A_o^{\operatorname{Spec}\mathfrak{A}}$ is continuous. Then $s^{-1}([a_i]\Theta_o)$ $=: O_i$ is a clopen set, $i = 1,\cdots,n$, and $\operatorname{Spec}\mathfrak{A} = O_1 \dot{\cup} \cdots \dot{\cup} O_n$. Gluing $a_1,\cdots, a_n$ together over $O_1,\cdots, O_n$ yields an $a \in A$ with $\hat{\hat{a}} = s$. ●

One immediate consequence of Theorem 16 is a positive answer to the Mycielski question for (D)-discriminator varieties.

<u>8.18. Corollary</u>. If $\mathfrak{A} \in \mathbb{V}$, where $\mathbb{V}$ is a (D)-discriminator variety, then $\mathfrak{A}$ is equationally compact iff $\mathfrak{A}$ is a retract of a compact algebra (in fact a product of finite algebras).

<u>proof</u>: Use Taylor's Theorem (8.12), Weglorz' Theorem (2.3), Theorem 16, and the fact that a product of retracts of compact algebras is again a retract of a compact algebra. ●

<u>8.19. Corollary</u>. Let $\mathbb{V}$ be a (D)-discriminator variety. For $\mathfrak{A} \in \mathbb{V}$ the following are equivalent:

(i) $\mathfrak{A}$ is a subdirect product of finite simple algebras.

(ii) $\mathfrak{A}$ is a subalgebra of a compact algebra.

(iii) $\mathfrak{A}$ is a subalgebra of an equationally compact algebra.

Corollary 19 implies in particular that an infinite simple algebra in $\mathbf{V}$ is not compactifiable; one can say even more, namely that such an algebra is not even quasi-compactifiable (cp. §4). Indeed, suppose $\mathcal{B}$ is a quasi-compactification of the infinite discriminator algebra $\mathfrak{A}$ (by Proposition 2.12 we may assume $\mathcal{B} \in \mathbf{HSP}(\mathfrak{A})$). Consider the system of equations

$$\Sigma = \{t(x_i,x_j,a) = x_i = t(x_i,x_j,b) \ ; \ i,j \in I, \ i \neq j\}$$

where $a,b \in A$, $a \neq b$, are fixed and I is a set larger than B . Σ is finitely solvable in $\mathfrak{A}$ because A is infinite, but Σ is not solvable in $\mathcal{B}$ due to the cardinality of I and the fact that $t(u,u,v) = v$ is an identity on $\mathcal{B}$.

A Boolean space is called <u>extremally disconnected</u> if the interior of every closed set is clopen. As a Boolean algebra is complete iff its Stone space is extremally disconnected, we also have at once:

<u>8.20. Corollary</u>. An m-ring R is equationally compact iff it is isomorphic to a finite sum of Boolean extensions of Galois fields by complete Boolean algebras iff it is isomorphic to a finite sum of rings each of which is the ring of continuous functions of an extremally disconnected Boolean Space into a finite field. In particular, an equationally compact m-ring has an identity.

We conclude the section with the promised characterization of the compact members of $\mathbf{V}$; with the description of the equationally compact ones most of the work has been done.

<u>8.21. Theorem</u>. Let $\mathbb{V}$ be a (D)-discriminator variety and suppose $(\mathfrak{A}, \mathcal{T})$ is a compact topological algebra with $\mathfrak{A} \in \mathbb{V}$. Then $(\mathfrak{A}, \mathcal{T})$ is iso- and homeomorphic to a product of finite simple algebras endowed with the Tychonoff product topology, and $\mathcal{T}$ is the only compact topology $\mathfrak{A}$ can carry.

<u>proof</u>: As $\mathfrak{A}$ is equationally compact Lemma 17 gives

$$\mathfrak{A} \cong \prod(\mathfrak{C}_i;\ i \in I) \quad \text{where} \quad \mathfrak{C}_i = \mathfrak{A}/\Theta(r_i,s_i) \cong \mathfrak{A}_i[B_i] ,$$

B_i being the complete Boolean algebra of clopen sets on $\operatorname{Spec}\mathfrak{C}_i$. To prove the algebraic statement of the theorem it will suffice to show, for each i, that B_i is isomorphic to a power set Boolean algebra.

Fix i arbitrary. As $\nabla \in P(\mathfrak{A})$ there are elements $0,1 \in A$ such that $\nabla\ (= A\times A) = \Theta(0,1)$. Setting $1_i :=$ $n(r_i,s_i,1,0)$ it follows that $E(r_i,s_i) = E(1,1_i)$. Now define

$$\begin{aligned}\hat{B}_i := \{a \in A;\ a \equiv 0(\phi) \text{ or } a \equiv 1(\phi) & \text{ for } \phi \in E(1,1_i) ,\\ \text{and } a \equiv 0(\phi) & \text{ for } \phi \in D(1,1_i)\} .\end{aligned}$$

As $\mathfrak{C}_i = \mathfrak{A}/\Theta(1,1_i)$ the elements of $\operatorname{Spec}\mathfrak{C}_i$ are of the form $\bar{\phi} := \phi/\Theta(1,1_i)$ with $\phi \in E(1,1_i)$. Also for brevity we denote the elements $[x]\Theta(1,1_i)$ $(x \in A)$ of $\mathfrak{C}_i$ by $\bar{x}$. Now define

$$\begin{aligned} f : \hat{B}_i &\longrightarrow B_i \\ x &\longmapsto E(\bar{1},\bar{x}) \quad (= \{\bar{\phi}\ ;\ \phi \in E(1,x)\}) ,\end{aligned}$$

and observe:

(1) f is injective: trivial verification.

(2) f is surjective: for arbitrary $\bar{u},\bar{v} \in S_i$ $n(u,v,1_i,0) \in \hat{B}_i$ and $E(\bar{u},\bar{v}) = E(\bar{1},\overline{n(u,v,1_i,0)})$.

(3) $f(1_i) = E(\bar{1},\bar{1}_i) = \operatorname{Spec}\mathfrak{C}_i$, $f(0) = E(\bar{1},\bar{0}) = \emptyset$.

Note further that for any $\phi \in E(1,1_i)$ and $x,y \in \hat{B}_i$,

$x \equiv 1(\phi)$ and $y \equiv 1(\phi)$ iff $n(x,y,x,0) \equiv 1(\phi)$ and

$x \equiv 1(\phi)$ or $y \equiv 1(\phi)$ iff $t(x,0,y) \equiv 1(\phi)$, hence

(4) $E(\bar{1},\bar{x}) \cap E(\bar{1},\bar{y}) = E(\bar{1},\overline{n(x,y,x,0)})$, $x,y \in \hat{B}_i$.

(5) $E(\bar{1},\bar{x}) \cup E(\bar{1},\bar{y}) = E(\bar{1},\overline{t(x,0,y)})$, $x,y \in \hat{B}_i$.

If we now define the operations $\wedge$ and $\vee$ on $\hat{B}_i$ by

$$x \wedge y := n(x,y,x,0) \quad (\in \hat{B}_i\ !)$$

$$x \vee y := t(x,0,y) \quad (\in \hat{B}_i\ !)$$

it follows from (1) - (5) that f^{-1} is a 0,1 - lattice isomorphism from B_i onto $\hat{B}_i = \langle \hat{B}_i ; \wedge, \vee, 0, 1_i \rangle$ and therefore $\hat{B}_i$ is itself a Boolean algebra isomorphic to B_i , where the complementation operation is given by

$$x' = n(x,1_i,0,1_i) .$$

As all operations on the Boolean algebra $\hat{B}_i$ are polynomials on $\mathfrak{A}$, $\hat{B}_i$ becomes a topological Boolean algebra with the relative topology inherited from A . Moreover, the elements of $\hat{B}_i$ are precisely the solutions in $\mathfrak{A}$ of the equation

$$x = n(1,x,1_i,0)$$

and hence form a closed subset of A . Thus $\hat{B}_i$ is a compact Boolean algebra, and therefore so is B_i , hence, as is well-known, B_i must be isomorphic to a power set Boolean algebra. This proves the algebraic statement. We have therefore an (algebraic) isomorphism

$$\mathfrak{A} \cong \prod(\mathfrak{L}_j \ ; j \in J) ,$$

where the $\mathfrak{L}_j$'s are finite simple algebras. The remaining statement of the theorem will follow if we show that $\mathfrak{A}$ can carry only one compact topology. As this property is an algebraic invariant, it suffices to show that the product topology on $\prod(C_j;\ j \in J)$ is the only compact topology the algebra $\prod(\mathcal{L}_j;\ j \in J)$ can carry. This is easily accomplished, for note that a subbasis for the closed sets of the product topology is given by the family of sets

$$S_{i,c_i} = \prod(C_j;\ j \in J,\ j \neq i) \times \{c_i\}\ , \qquad i \in J,\quad c_i \in C_i\ .$$

For a fixed S_{i,c_i} choose $a \in \prod(C_j;\ j \in J)$ with $a(i) = c_i$, and define $c \in \prod(C_j;\ j \in J)$ by

$$c(j) = \begin{cases} c_i & \text{if } i = j \\ a(j) & \text{if } i \neq j\ . \end{cases}$$

Then S_{i,c_i} is the solution set of the equation

$$t(c,x,a) = a\ ,$$

hence S_{i,c_i} is a closed set in <u>any</u> compact topology defined on the algebra $\prod(\mathcal{L}_j;\ j \in J)$. Therefore the product topology is coarser than all other compact topologies, hence the unique one. ●

<u>8.22. Corollary</u>. A compact m-ring is isomorphic to a product of Galois fields, and the compact topology it can carry is unique.

The algebraic part of Corollary 22 follows of course from [8, Theorem 4.1], as arithmetic varieties of rings are quasi-primal. Since m-rings are semisimple it is also subsumed, from the ring theoretic side, by even earlier results: I. Kaplansky proved in 1947 in one of his ground-breaking papers on topological rings the following structure theorem for compact semisimple rings [28, Theorem 16] : a compact topological semisimple ring is iso- and homeomorphic to a product of finite simple rings; much later, in 1960, S. Warner published a proof of the fact that a semisimple ring can carry only one compact topology [51, Theorem 2]. His argument on the surface appears to require the full strength of Kaplansky's structure theorem. That this is not the case, that in fact the uniqueness of topology property follows rather quickly from an even weaker statement than Kaplansky's theorem, is evident from the following observation, with which we conclude the paragraph.

8.23. Proposition. If R_i , $i \in I$, are finite rings with identity then the ring $R = \prod(R_i;\ i \in I)$ carries exactly one compact topology, namely the product topology.

proof: As in the proof of Theorem 21 we need only verify that a typical subbasic closed set (in the product topology) of the form

$$S_{i,c_i} = \prod(R_j;\ j \in J,\ j \neq i) \times \{c_i\} , \quad i \in I, \quad c_i \in R_i ,$$

is closed in any compact topology defined on the ring R , by illustrating an equation of which it is the solution set. If for $a \in R_i$ $\hat{a}$ denotes the element of R defined by

$$\hat{a}(j) = \delta_{ij} \cdot a , \quad j \in I ,$$

and if e_i is the identity of R_i , then the equation

$$x \cdot \hat{e}_i = \hat{c}_i$$

does the trick. ●

§9 Injectivity

In this section we exploit the theory developed in the previous paragraphs to relate injectivity and equational compactness. There are no non-trivial injective rings, the reader may remark, and indeed, the only injective in the category of unital rings, unital commutative rings, etc. (see R.Raphael [42]) is the trivial ring (0) . On the other hand, if the category in question is suitably restricted, thereby allowing the possibility of more injectives, there is no inherent loss of interest in the injective objects in the smaller category, as long, of course, as the category remains nice - for example, as long as it is still an equational class. (See B. Banaschewski [2] and A. Day [10] for the basic theory of injectivity in equational classes.) Now every algebra $\mathfrak{A}$ in a category of algebras $\mathbb{K}$ which is injective in $\mathbb{K}$ will be equationally compact as long as $\mathbb{K}$ contains at least $\mathbb{HSP}(\mathfrak{A})$; this is because injectivity implies pure-injectivity, and the pure-injectivity of an algebra $\mathfrak{A}$ in $\mathbb{HSP}(\mathfrak{A})$ insures already equational compactness. Indeed, an algebra $\mathfrak{A} \in \mathbb{K}(\tau)$ is pure-injective in $\mathbb{K}(\tau)$ iff it is pure-injective in $\mathbb{HSP}(\mathfrak{A})$, as a close inspection of Weglorz' Theorem (2.3) reveals.

Therefore the question arises naturally whether or not we can "approximate" the non-categorical property of equational compactness via the categorical property of injectivity with respect, of course, to an appropriately diminished variety.

By way of illustration let us survey the situation in the equational class $\mathbb{G}$ of abelian groups. The injective objects in $\mathbb{G}$ are precisely the divisible groups (which exclude the finite ones, all of which are equationally compact!), and, recalling the well-known argument, divisibility is deduced from injectivity by completing diagrams containing the given group and subgroups of $\mathbb{Z}^+$. This is not too surprising, however, since $\mathbf{G} = \mathbb{HSP}(\mathbb{Z}^+)$. By this observation, too, it is clear that a proper subvariety $\mathbf{V}$ of $\mathbf{G}$ will have to exclude $\mathbb{Z}^+$ and this in turn requires admitting a torsion equation $m \cdot x = 0$ as an identity on $\mathbf{V}$. If moreover the equationally compact members of $\mathbf{V}$ are to be injective, then $\mathbf{V}$ must satisfy $m \cdot x = 0$ where m is square-free. (Since the cyclic group of order p^2, p prime, does not retract onto its subgroup of order p, the latter is equationally compact but would not be injective in $\mathbf{V}$ if $\mathbf{V}$ contained the former.) Thus $\mathbf{V}$ must consist of modules over the ring $R = GF(p_1) \oplus \cdots \oplus GF(p_n)$ for mutually distinct primes $p_1, \cdots, p_n$, so now <u>every</u> group in $\mathbf{V}$ is injective in $\mathbf{V}$ as R is semisimple. In summary:

<u>9.1. Proposition</u>. A variety $\mathbf{V}$ of abelian groups, alias zero rings, has the property that every equationally compact object is injective in $\mathbf{V}$ iff $\mathbf{V}$ satisfies an identity $m \cdot x = 0$ with square-free m.

Turning now to arbitrary rings, the circumstances are more involved, and we only give an analysis within the framework covered by our investigations of discriminator varieties. Restricting to these varieties it is possible to do even more, namely to <u>characterize</u> injectivity in these classes. The special cases where injectivity and equational compactness coincide are a by-product. That there are other than just "trivial" cases - in the sense illustrated by abelian groups - is supported by a theorem of B Weglorz [53, Theorem 4.1] which has already been referred to (the equivalence of (ii) and (iii) was proved first by P.R. Halmos [23]) :

9.2. Proposition (B. Weglorz [53]). For the class $\mathbb{B}$ of Boolean algebras the following are equivalent for $B \in \mathbb{B}$:

(i) B is equationally compact.

(ii) B is complete.

(iii) B is injective in $\mathbb{B}$.

$\mathbb{B}$ of course "is" the variety of unital 2-rings (alias Boolean rings) as we do not distinguish between reduct-equivalents. In this paragraph the above theorem of Weglorz is put in its natural setting - namely, by characterizing those discriminator varieties in which equational compactness and injectivity coincide. Weglorz' result will then fall out as an example. The equivalence of (ii) and (iii) above has been generalized by D. Haines [17] to varieties generated by a single finite prime field, where, of course, the "completeness" concept is suitably extended. We note that the ring theoretic results presented here follow also from the characterization of injectives in quasi-primal varieties in H. Werner [59] (the necessary translation into our formulation is rendered by a characterization of demi-semi-primal algebras due to R. Quackenbush; see [59]) and also have a proper intersection (the $\mathbb{V} = \mathbb{V}(GF(p))$ case) with Theorem 5.8 of A. Day [10] concerning injectivity in certain equationally complete varieties. See also W. Taylor [48] in this context.

Let us make a general definition:

9.3. Definition. Let $\mathbb{K}$ be an equational class. An algebra $\mathfrak{A}$ in $\mathbb{K}$ is residually injective in $\mathbb{K}$ iff every diagram

$$\begin{array}{ccc} \mathfrak{L} & & \\ \uparrow & & \\ \mathfrak{D} & \rightarrow & \mathfrak{A} \end{array}$$

with $\mathfrak{L}$ and $\mathfrak{D}$ subdirectly irreducible members of $\mathbb{K}$ has a commutative completion $\mathfrak{L} \rightarrow \mathfrak{A}$.

Obviously the more transparent the class of subdirectly irreducibles is, the easier it is to check for residual injectivity, and thus under such circumstances it would be of value to characterize injectivity in terms of residual injectivity. The extent to which this is possible in discriminator varieties is the content of the main theorem, 9.5. We begin with an easy lemma, which is actually implicit in the proof of Theorem 4.1 of A. Day [10].

9.4. Lemma. If $\mathfrak{A}$ is simple and residually injective in the discriminator variety $\mathbb{V}$, then $\mathfrak{A}$ is injective in $\mathbb{V}(\mathfrak{A})$.

proof: Since every $\mathfrak{B}$ in $\mathbb{V}(\mathfrak{A})$ is a subalgebra of a power of $\mathfrak{A}$, it suffices to complete diagrams of the following type:

$$\begin{array}{ccc} \mathfrak{L} & & \\ \cup\!\! \mid & & \\ \mathfrak{B} & \xrightarrow{f} & \mathfrak{A} \end{array}$$

where $\mathfrak{L} = \mathfrak{A}^I$ for an index set I . We may assume that $|f(B)| > 1$, since otherwise f lifts trivially. As $f(\mathfrak{B})$ is simple there is an ultrafilter F on I (Lemma 8.7) such that $\ker(f) = \Theta_F|_B$. This yields the diagram

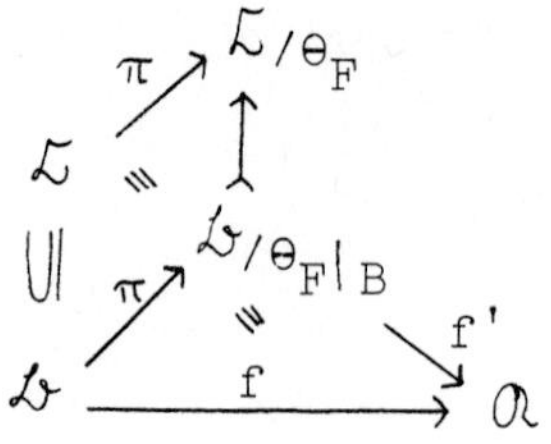

Since $\mathfrak{B}/\theta_{F|B}$ and $\mathfrak{L}/\theta_F$ (= $\mathfrak{A}_F^I$) are simple, residual injectivity of $\mathfrak{A}$ guarantees the existence of an extension $g : \mathfrak{L}/\theta_F \to \mathfrak{A}$ of f'. $g \circ \pi$ is then the desired extension of f . ●

<u>9.5. Theorem</u>. Let $\mathbf{V}$ be a (D)-discriminator variety. For $\mathfrak{A}$ in $\mathbf{V}$ the following are equivalent:

(i) $\mathfrak{A}$ is injective in $\mathbf{V}$.

(ii) $\mathfrak{A}$ is equationally compact and every finite simple quotient of $\mathfrak{A}$ is residually injective in $\mathbf{V}$.

(iii) $\mathfrak{A} \cong \prod(\mathfrak{A}_i[B_i] ; i \in I)$, where the B_i's are complete Boolean algebras and the $\mathfrak{A}_i$'s are pairwise non-isomorphic, finite, simple, and residually injective in $\mathbf{V}$.

<u>proof</u>: (ii) $\Leftrightarrow$ (iii): In light of Theorem 8.16 we may assume that $\mathfrak{A} = \prod(\mathfrak{A}_i[B_i] ; i \in I)$ with B_i complete and the $\mathfrak{A}_i$'s finite simple and pairwise non-isomorphic, and need only show that the proper finite simple quotients of $\mathfrak{A}$ are given up to isomorphism by the family $\{\mathfrak{A}_i ; i \in I\}$. Obviously each $\mathfrak{A}_i$ is a quotient of $\mathfrak{A}$. Conversely, let $\mathfrak{L}$ be finite simple and suppose that $f : \prod(\mathfrak{A}_i[B_i] ; i \in I) \to \mathfrak{L}$ is an epimorphism. For each $i \in I$ there is a set J_i such that $\mathfrak{A}_i[B_i]$ is a subdirect product of $\mathfrak{A}_i^{J_i}$ containing the diagonal. By Lemma 8.7

$$\ker(f) \;=\; \theta_F \mid \prod(A_i[B_i];\, i\in I)$$

for some ultrafilter F on $\dot{\bigcup}(J_i\ ;\ i\in I)$. It follows by Loś' Theorem (1.1) that $\bigcup(J_i\ ;\ |A_i| = |C|) \in F$. By condition (D) only a finite number of $\mathfrak{A}_i$'s satisfy $|A_i| = |C|$, and hence $J_{i_o} \in F$ for some i_o , whence

$$\begin{aligned}\mathfrak{C} \;\cong\; \prod(\mathfrak{A}_i[B_i]\,;\, i\in I)/\ker(f) \;&\cong\; \mathfrak{A}_{i_o}[B_{i_o}]/\theta_{F'} \mid A_i[B_i] \\ \subseteq\; \mathfrak{A}_{i_o}^{J_{i_o}}/\theta_{F'} \;&\cong\; \mathfrak{A}_{i_o}\end{aligned}$$

where F' is the restriction of F onto J_{i_o}. But $\theta_{F'}$ separates the diagonal elements in $A_{i_o}^{J_{i_o}}$ and hence the inclusion above is equality. Thus $\mathfrak{C} \cong \mathfrak{A}_{i_o}$ as desired.

(i) $\Rightarrow$ (iii): Let $\mathfrak{A}$ be injective. Then $\mathfrak{A}$ is equationally compact and

$$\mathfrak{A} \;\cong\; \prod(\mathfrak{A}_i[B_i]\,;\, i\in I)\ ,$$

where the B_i's are complete and the $\mathfrak{A}_i$'s are finite simple and pairwise non-isomorphic (Theorem 8.16). To show is that each $\mathfrak{A}_i$ is residually injective. Now $\mathfrak{A}_i$ is a retract of $\mathfrak{A}_i[B_i]$ as the latter contains the diagonal and thus $\mathfrak{A}' \;:=\; \prod(\mathfrak{A}_i\ ;\ i\in I)$ is a retract of $\mathfrak{A}$, hence injective itself. Now suppose $\mathfrak{D}$ and $\mathfrak{C}$ are simple, $\mathfrak{D}\subseteq\mathfrak{C}$, and $f : \mathfrak{D} \to \mathfrak{A}_i$ is a homomorphism. If $|f(D)| = 1$ then f lifts trivially to $\mathfrak{C}$. So assume $|f(D)| \geq 2$. Setting $\mathfrak{D}' \;:=\; \mathfrak{D}\times\prod(\mathfrak{A}_j\ ;\ j\neq i)$ and $\mathfrak{C}' \;:=\; \mathfrak{C}\times\prod(\mathfrak{A}_j\ ;\ j\neq i)$ we have a naturally defined homomorphism $f' : \mathfrak{D}' \to \mathfrak{A}'$, and by injectivity there is a commutative completion

$$\begin{array}{lll} \mathfrak{C}' & \overset{\bar{f}'}{\searrow} & \\ \cup\!\!\mid & & \\ \mathfrak{D}' & \xrightarrow{\ f'\ } & \mathfrak{A}'\ . \end{array}$$

If $\pi_i : \mathfrak{A}' \to \mathfrak{A}_i$ is the canonical projection then $\pi_i \circ \bar{f}'(\mathcal{L}')$ is finite simple and non-trivial; indexing the factor $\mathcal{L}$ in the product $\mathcal{L}'$ by i, Lemma 8.7 yields an ultrafilter F on I such that $\ker(\pi_i \circ \bar{f}') = \Theta_F$. Exploiting (D) and Loś' Theorem again it follows that F is principal, as the $\mathfrak{A}_i$'s were chosen pairwise non-isomorphic. But then F is generated by $\{i\}$, as $\bar{f}'$ does not collapse $\mathcal{L}$. Hence there is a commutative extension of the above diagram to

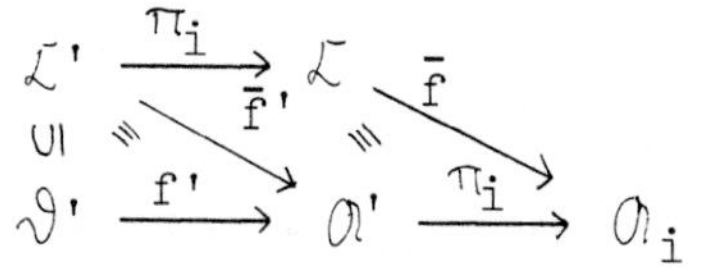

and $\bar{f}$ is the desired homomorphism extending f.

(iii) $\Rightarrow$ (i): Let $\mathfrak{A} = \prod(\mathfrak{A}_i[B_i]\,;\ i \in I)$ be as in (iii). We have the following diagram to complete:

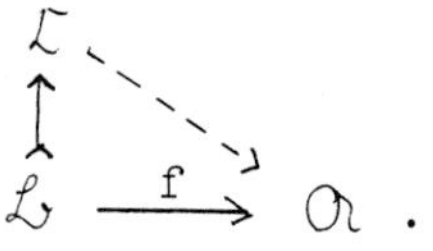

Without loss of generality we may assume there is a set $X = \{\mathfrak{A}_j;\ j \in J\}$ of pairwise non-isomorphic algebras such that

(1) $I \subseteq J$ (i.e., $\mathfrak{A}_i \in X$ for each $i \in I$),

(2) there is a product $\mathcal{T}$, indexed by a set K, whose factors are in X and such that $\mathcal{L}$ is a subalgebra of $\mathcal{T}$, and $\mathfrak{L}$ is a subalgebra of $\mathcal{L}$, and

(3) for all $i \in I$ and $j \in J$ $\mathfrak{A}_j$ is not isomorphic to any subalgebra of $\mathfrak{A}_i$.

(Note: (3) can be achieved without sacrificing (1) due to the fact that the $\mathfrak{A}_i$'s are residually injective for all $i \in I$.) For $j \in J$ let $K_j \subseteq K$ index those factors of $\mathcal{T}$ lying in $\mathbf{V}(\mathfrak{A}_j)$, and let π_j be the projection onto the product of these factors. (If $K_j = \emptyset$ let $\pi_j(\mathcal{C})$ be the one-element algebra.) Set $\mathcal{B}_j = \pi_j(\mathcal{B})$, $\mathcal{L}_j = \pi_j(\mathcal{L})$, and for $i \in I$ denote the projection of $\mathfrak{A}$ onto $\mathfrak{A}_i[B_i]$ also by π_i.

Fix $i \in I$. We claim there is a commutative completion $f_i : \mathcal{B}_i \to \mathfrak{A}_i[B_i]$ of the diagram

$$\begin{array}{ccc} \mathcal{B} & \xrightarrow{f} & \mathfrak{A} \\ \pi_i \downarrow & & \pi_i \downarrow \\ \mathcal{B}_i & & \mathfrak{A}_i[B_i]. \end{array}$$

Indeed, we <u>must</u> define f_i by $f_i(b_i) := \pi_i(f(b))$, where $b_i = \pi_i(b)$; f_i will be the desired homomorphism if it is a well-defined map.

Pick $b, d \in B$ such that $\pi_i(b) = b_i = \pi_i(d)$, and assume that $\pi_i(f(b)) \neq \pi_i(f(d))$. Let $p : \mathfrak{A}_i[B_i] \to \mathfrak{A}_i$ be a canonical projection satisfying $p \circ \pi_i(f(b)) \neq p \circ \pi_i(f(d))$. Set $g := p \circ \pi_i \circ f$. Then $g(\mathcal{B})$ is a (simple) subalgebra of $\mathfrak{A}_i$, say $\mathfrak{A}_i'$, with at least two elements. By Lemma 8.7 $\ker(g) = \theta_{F|B}$ for an ultrafilter F on K. Let $K' := \{k \in K ; b(k) \neq d(k)\}$. (The figure at right may serve as a mnemonic.) Since $\pi_i(b) = \pi_i(d)$ but b and d are not congruent modulo $\ker(g)$, it follows that

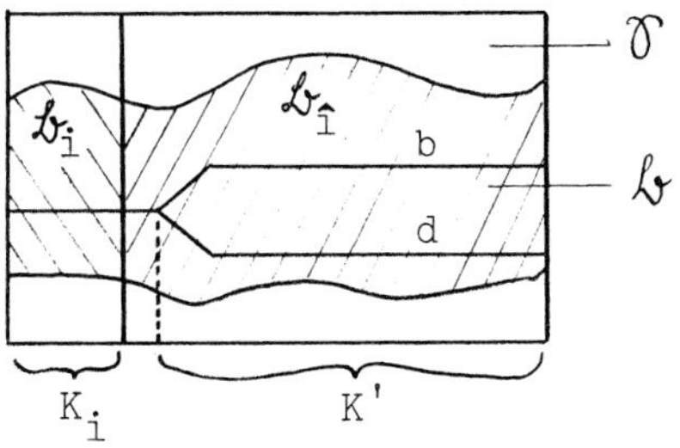

$\{k \in K ; b(k) = d(k)\} \notin F$. We conclude that $K' \subseteq K \setminus K_i$ and $K' \in F$, as F is an ultrafilter. Set $\mathcal{B}_{\hat{i}} := \prod(\mathcal{B}_j ; j \neq i)$ and let $\pi_{\hat{i}} : \mathcal{B} \to \mathcal{B}_{\hat{i}}$ be the canonical projection. Then there is a commutative completion $\bar{g} : \mathcal{B}_{\hat{i}} \to \mathfrak{A}_i$ of the diagram

$$\begin{array}{ccc} \mathfrak{B} & \xrightarrow{\;g\;} & \mathfrak{A}_i' \\ {\scriptstyle \pi_{\hat{i}}}\big\downarrow & & \\ \mathfrak{B}_{\hat{i}} & & \end{array}$$

because $\ker(\pi_{\hat{i}}) \subseteq \ker(g)$. Indeed, from $\pi_{\hat{i}}(x) = \pi_{\hat{i}}(y)$ it follows that $K_o := \{k \in K \; ; \; x(k) = y(k)\} \supseteq K \setminus K_i$ (condition (3) !) $\supseteq K' \in F$, hence $K_o \in F$, i.e., $x \equiv y \; (\Theta_F)$ which by construction implies $g(x) = g(y)$. Since $\mathfrak{B}_{\hat{i}}$ is in $\mathbf{V}(\mathfrak{A}_j ; \; j \in J, \; j \neq i)$ we conclude that $\bar{g}(\mathfrak{B}_{\hat{i}}) = \mathfrak{A}_i'$ is in $\mathbf{V}(\mathfrak{A}_j ; \; j \in J \;\; j \neq i)$ too. By Proposition 8.6 $\mathfrak{A}_i'$ is isomorphic to a subalgebra of an ultraproduct $\prod_D(\mathfrak{A}_j ; \; j \in J, \; j \neq i) =: \mathfrak{A}'$. Since $\mathfrak{A}_i$ is residually injective and $\mathfrak{A}'$ is simple there is a commutative completion of the diagram

$$\begin{array}{ccc} \mathfrak{A}' & & \\ \big\uparrow & & \\ \mathfrak{A}_i' & \subseteq & \mathfrak{A}_i \end{array}$$

embedding $\mathfrak{A}'$ into $\mathfrak{A}_i$, since $\mathfrak{A}_i'$ has at least two elements. In particular, $\mathfrak{A}'$ is finite, hence the usual argument with (D) and Łoś' Theorem gives that $\mathfrak{A}' \cong \mathfrak{A}_j$ for some $j \neq i$, i.e., $\mathfrak{A}_j$ is isomorphic to a subalgebra of $\mathfrak{A}_i$, contradicting (3). It follows that f_i is well-defined, which is all we needed.

The proof is now quickly completed. For each $i \in I$ we have the diagram

$$\begin{array}{ccc} \mathfrak{L}_i & & \\ \cup\!\mid & & \\ \mathfrak{B}_i & \xrightarrow{\;f_i\;} & \mathfrak{A}_i[B_i] \end{array}$$

with f_i as constructed above. Note that all occurring

algebras are in $\mathbb{V}(\mathfrak{A}_i)$. By Taylor's Theorem (8.12) the canonical embedding

$$\mathfrak{A}_i[B_i] \rightarrowtail \prod(\mathfrak{A}_i[B_i]/\theta \ ; \ \theta \in \mathrm{Spec}(\mathfrak{A}_i[B_i]))$$

is pure. By the same argument used in the proof of (ii) $\Leftrightarrow$ (iii) it follows that each $\mathfrak{A}_i[B_i]/\theta$ in the above product is isomorphic to $\mathfrak{A}_i$; hence $\mathfrak{A}_i[B_i]$ is a pure subalgebra of a power $\mathfrak{A}_i{}^L$, and the latter is injective in $\mathbb{V}(\mathfrak{A}_i)$ since $\mathfrak{A}_i$ enjoys this property (Lemma 4). $\mathfrak{A}_i[B_i]$ is equationally compact hence a retract of $\mathfrak{A}_i{}^L$ by Weglorz' Theorem (2.3), and therefore is itself injective in $\mathbb{V}(\mathfrak{A}_i)$. Thus we have a commutative completion $f_i' : \mathcal{L}_i \longrightarrow \mathfrak{A}_i[B_i]$ of the above diagram. Now define $f' : \mathcal{L} \longrightarrow \mathfrak{A}$ by.

$$f'(c) \ := \ (f_i'(c_i))_{i \in I} \in \prod(A_i[B_i] \ ; \ i \in I) \ = \ A \ ,$$

where $c_i = \pi_i(c) \in C_i$ for $i \in I$. That f' is the desired extension of f is checked by chasing the diagram

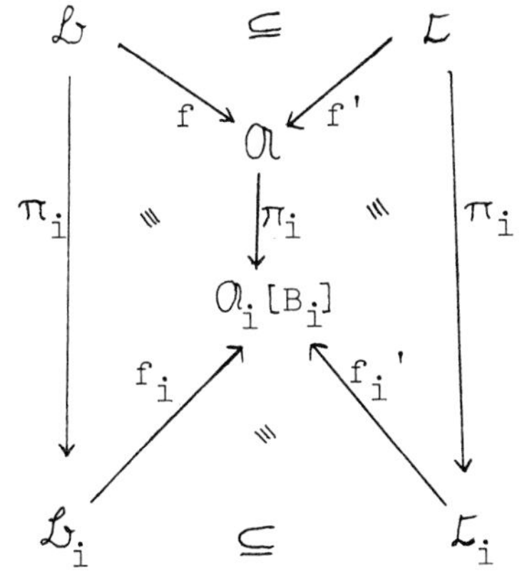

in a clockwise direction. $\bullet$

<u>9.6. Corollary</u>. Let $\mathbf{V}$ be a (D)-discriminator variety. Then every equationally compact algebra in $\mathbf{V}$ is injective in $\mathbf{V}$ if and only if no finite simple algebra in $\mathbf{V}$ with

more than one element is a proper subalgebra of a simple algebra in $\mathbb{V}$.

proof: Indeed, the latter condition holds in $\mathbb{V}$ iff every finite simple algebra is residually injective in $\mathbb{V}$. •

Following the terminology of B. Banaschewski [2] we say that an equational class $\mathbb{K}$ has enough injectives if every $\mathfrak{A}$ in $\mathbb{K}$ has an injective extension in $\mathbb{K}$, i.e., $\mathfrak{A} \subseteq \mathfrak{B}$ with $\mathfrak{B}$ injective in $\mathbb{K}$. If, moreover, $\mathfrak{B}$ is an essential extension of $\mathfrak{A}$ (i.e., the only congruence $\Theta \in C(\mathfrak{B})$ with $\Theta|_A = id_A$ is id_B), then $\mathfrak{B}$ is an injective hull of $\mathfrak{A}$. A criterium for the existence of injective hulls is given by

9.7. Proposition (B. Banaschewski [2]). For an equational class $\mathbb{K}$ the following are equivalent:

(i) $\mathbb{K}$ has enough injectives.

(ii) Every $\mathfrak{A}$ in $\mathbb{K}$ has an injective hull.

One of the consequences of the conditions in 9.7 is that every $\mathfrak{A}$ in $\mathbb{K}$ must have, up to isomorphism, only a set of essential extensions. In a discriminator variety this generally fails, as any infinite simple algebra has arbitrarily large ultrapowers, and these are essential extensions (being simple themselves). Thus we must restrict our attention to quasi-primal varieties if we want to hope to find enough injective hulls. In the following we in fact sharpen 9.7 for quasi-primal varieties.

9.8. Theorem. Let $\mathbb{V}$ be a quasi-primal variety. For $\mathfrak{A}$ in

$\mathbf{V}$ the following are equivalent:

(i) Every simple quotient of $\mathfrak{A}$ is contained in a residually injective simple algebra.

(ii) $\mathfrak{A}$ has an injective extension.

(iii) $\mathfrak{A}$ has an injective hull.

<u>proof</u>: (i) $\Rightarrow$ (ii): We have

$$\mathfrak{A} \rightarrowtail \prod(\mathfrak{A}/\theta;\ \theta \in \operatorname{Spec}\mathfrak{A}) \subseteq \prod(\mathfrak{C}_\theta;\ \theta \in \operatorname{Spec}\mathfrak{A}) =: \mathfrak{B}$$

where for $\theta \in \operatorname{Spec}\mathfrak{A}$ $\mathfrak{C}_\theta$ is a simple residually injective algebra containing $\mathfrak{A}/\theta$. $\mathfrak{B}$ is injective as it obviously satisfies (iii) of Theorem 5.

(ii) $\Rightarrow$ (i): If $\mathfrak{B}$ is an injective extension of $\mathfrak{A}$ then we have

$$\mathfrak{A} \subseteq \mathfrak{B} \rightarrowtail \prod(\mathfrak{B}/\theta\ ;\ \theta \in \operatorname{Spec}\mathfrak{B})$$

where each $\mathfrak{B}/\theta$ is residually injective. A direct application of Lemma 8.7 yields that any simple quotient of $\mathfrak{A}$ is isomorphic to a subalgebra of some $\mathfrak{A}_i$.

(ii) $\Leftrightarrow$ (iii): Observe that, for residually injective simple algebras $\mathfrak{A}_1,\cdots,\mathfrak{A}_n$ in $\mathbf{V}$, $\mathfrak{A}_1,\cdots,\mathfrak{A}_n$ are <u>precisely</u> the residually injective simple members of $\mathbf{V}' := \mathbf{V}(\mathfrak{A}_1,\cdots,\mathfrak{A}_n)$. It follows from Theorem 5 (ii) that, for $\mathfrak{B}$ in $\mathbf{V}$, $\mathfrak{B}$ is injective in $\mathbf{V}'$ iff $\mathfrak{B}$ is injective in $\mathbf{V}$. Thus, if $\mathfrak{A}$ in $\mathbf{V}$ has an injective extension $\mathfrak{B}$, and if $\mathfrak{A}_1,\cdots,\mathfrak{A}_n$ are the (residually injective!) simple quotients of $\mathfrak{B}$, then $\mathfrak{A}$ has an injective extension in $\mathbf{V}' = \mathbf{V}(\mathfrak{A}_1,\cdots,\mathfrak{A}_n)$. But $\mathbf{V}'$ has enough injectives by Proposition 8.6 and the implication (i) $\Rightarrow$ (ii). Thus 9.7 applies, yielding an injective hull of $\mathfrak{A}$ in $\mathbf{V}'$ which, by the above observation is an injective hull in $\mathbf{V}$. ●

<u>9.9. Corollary</u>. A quasi-primal variety has enough injectives iff every simple member is a subalgebra of a residually injective simple member.

Let us, in conclusion, examine the precipitate of our general results when interpreted in the setting of m-rings.

If $\mathbf{V}$ is an arithmetic variety of rings, call a Galois field $GF(p^n)$ in V <u>maximal in $\mathbf{V}$</u> if every field in $\mathbf{V}$ with characteristic p is isomorphic to a subfield of $GF(p^n)$. If $GF(p^n)$ is residually injective in $\mathbf{V}$ then it is maximal in $\mathbf{V}$. Indeed, if $GF(p^m)$ is also in $\mathbf{V}$, then the diagram

$$\begin{array}{ccc} GF(p^m) & & \\ \uparrow & & \\ GF(p) & \rightarrowtail & GF(p^n) \end{array}$$

has a commutative completion if $GF(p^n)$ is residually injective, thus embedding $GF(p^m)$ into $GF(p^n)$. Conversely, suppose $GF(p^n)$ is maximal in $\mathbf{V}$, and let the diagram

$$\begin{array}{ccc} S & & \\ \uparrow & & \\ R & \xrightarrow{f} & GF(p^n) \end{array}$$

be given, with R and S fields and f a homomorphism such that $|f(R)| > 1$. Then f is an embedding, so $R = GF(p^r)$ for some r and hence $S = GF(p^s)$ for some $s \geq r$; we may assume that $R \subseteq S$. It follows that S is a splitting field over R of the polynomial $x^{p^s} - x$. Since $GF(p^n)$ is maximal in $\mathbf{V}$ it contains a subfield S' isomorphic to S, hence S' is a splitting field of $x^{p^s} - x$ over $f(R)$. ($f(R)$ is contained in S' as Galois fields contain at most one subfield of a pregiven order.) By a fundamental result on splitting fields it follows that there is an isomorphism $f' : S \to S'$ extending f. We conclude that $GF(p^n)$ is residually injective in $\mathbb{V}$. These short deliberations yield the following adaptation of Theorem 5 :

9.10. Theorem. Let V be a variety of m-rings. Then for R in $\mathbb{V}$ the following are equivalent:

(i) R is injective in $\mathbb{V}$.

(ii) R is equationally compact and for every maximal ideal $\mathfrak{a}$, $R/\mathfrak{a}$ is maximal in $\mathbb{V}$.

(iii) $R \cong F_1[B_1] \oplus \cdots \oplus F_n[B_n]$, where the B_i's are complete Boolean algebras and the F_i's are Galois fields, all maximal in $\mathbb{V}$.

Proofs of the following consequences of the foregoing are now immediate.

9.11. Corollary. In a variety of m-rings V, equational compactness and injectivity coincide if and only if $\mathbb{V}$ is generated by prime fields.

9.12. Corollary. A variety of m-rings $\mathbb{V}$ has enough injectives if and only if $\mathbb{V}$ is generated by Galois fields $GF(p_1^{n_1}), \ldots, GF(p_s^{n_s})$ for distinct primes $p_1, \ldots, p_s$.

A slightly different form of Corollary 12 was proved by B. Banaschewski in [2], There it is shown that a class of unital rings generated by a finite number of Galois fields of different characteristics has enough injectives. (See 9.15 below, where the apparent dissonance is clarified.)

9.13. Corollary. In a variety of m-rings $\mathbf{V}$, R in $\mathbf{V}$ has an injective hull in $\mathbf{V}$ iff for every maximal ideal $\mathfrak{a}$, $R/\mathfrak{a}$ is contained in a Galois field which is maximal in $\mathbf{V}$.

Note that the conditions of 9.12 and 9.13 are not empty. Consider, for example, the variety $\mathbf{V}$ generated by $GF(p^2)$ and $GF(p^3)$ for a fixed prime p . The fields in $\mathbf{V}$ are precisely $GF(p)$, $GF(p^2)$ and $GF(p^3)$, and none of these are maximal in $\mathbf{V}$. ($GF(p^2)$ satisfies the identity $x^{p^2} = x$ but not the identity $x^{p^3} = x$ satisfied by $GF(p^3)$, so the former is not a subfield of the latter.) Thus we have

9.14. Example. Let p and q be distinct primes.

(1) $\mathbf{V}(GF(p^2), GF(p^3))$ has no (non-trivial) injectives.

(2) $\mathbf{V}(GF(p^2), GF(p^3), GF(q))$ has some (non-trivial) injectives but not enough.

We conclude the section and the chapter by clarifying the connection between "ring" injectivity and "unital ring" injectivity for m-rings.

9.15. Theorem. Let $F_1, \cdots, F_n$ be Galois fields. Let $\mathbf{V}$ be the variety of rings generated by $F_1, \cdots, F_n$ and let $\mathbf{V}_1$ be the variety of unital rings generated by the unital rings $(F_1)_1, \cdots, (F_n)_1$ (where morphisms in $\mathbf{V}_1$ are of course unital ring homomorphisms). Let $R \in \mathbf{V}$. Then

R is injective in $\mathbb{V}$ if and only if R has an identity and R_1 is injective in $\mathbb{V}_1$.

<u>proof</u>: By Corollary 8.20 the equational compactness of R insures that R has an identity. Now both $\mathbb{V}$ and $\mathbb{V}_1$ are discriminator varieties, and the simple members of $\mathbb{V}$ and $\mathbb{V}_1$ coincide, modulo the discrepancy of type. Moreover, if F is a field in $\mathbb{V}$, then F is residually injective in $\mathbb{V}$ iff F_1 is residually injective in $\mathbb{V}_1$, simply because ring homomorphisms between fields which are not trivial are unital ring homomorphisms. It follows then immediately that condition (ii) of Theorem 5 holds for $R \in \mathbb{V}$ iff R has an identity and (ii) holds for R_1 in $\mathbb{V}_1$. The statement then follows by the equivalence of (i) and (ii) in Theorem 5. •

<u>9.16. Corollary</u>. If R is an m-ring with identity and R_1 is injective in $\mathbb{V}(R_1)$ then R is injective in $\mathbb{V}(R)$.

This section has dealt, in particular, with the question of injectivity in certain varieties of rings; specifically, those arithmetic varieties in which injectivity and equational compactness coincide have been characterized. A natural question is lurking in the shadows of all this: Are these <u>all</u> the equational classes of rings (apart from the zero ring case) satisfying this property? Any variety of rings which enjoys this property contains only a finite number of fields (which are then finite prime fields), a typical characteristic of arithmetic varieties. Indeed, suppose $\mathbb{V}$ is such a variety and suppose $\mathbb{V}$ contained $\mathbb{Q}$. Then $\mathbb{V}$ contains $\mathbb{Q}_p^*$, the field of p-adic numbers, since a topological ring satisfies every identity satisfied by a dense subring. But then $\mathbb{Z}_p^*$ is in $\mathbb{V}$, is equationally compact, but of course not retract of its quotient field $\mathbb{Q}_p^*$, a contradiction. Since $\mathbf{V}$ can contain no proper extensions of a finite prime field, the only other way to contain infinitely many fields is for it to contain infinitely many prime fields. But then any proper ultraproduct of all these is a field of characteristic zero, hence $\mathbb{Q} \in \mathbb{V}$ and we are back where we started. This observation is not discouraging, but lacking a sufficiently sophisticated structure theory for subdirectly irreducible rings (even commutative ones), the prospects of superficially extending the results of §9 to more general varieties, even to commutative unital varieties, appear rather dim, and any conjecture one way or the other seems overly bold.

CHAPTER V.
THE MYCIELSKI QUESTION

In the foregoing the strong association between equational compactness and compactness has become more than apparent. And thus the quest for a solution to the Mycielski Question in the class of <u>all</u> rings (in the subclasses studied so far the answer has been positive) is of particular interest. Positive answers to the question hold in the class of unital R-modules for any unital ring R (R.B. Warfield, Jr. [50]), the class of semilattices (S. Bulman-Fleming [7]) and the class of mono-unary algebras (G.H. Wenzel [56]), not to mention the various classes of rings and discriminator varieties dealt with in the discussions in the previous chapters.

The intent of this short and final chapter is to briefly survey the situation with regard to <u>negative</u> results re the Mycielski Question. Although we cannot give a definitive answer to the question in the class of associative rings there are indications that a negative answer might well be conjectured. Namely, there exists a counterexample to the Mycielski Question which possesses, as reduct, an abelian group C and which itself is a reduct of a ring A with $A^+ = C$. This is the main result of the chapter.

At the core of the matter lies a counterexample in the class of graphs due to W. Taylor, and so we begin §10 with the necessary graph theoretical background.

§10 Graphs and W. Taylor's examples

An algebraic system system $G = \langle G; R\rangle$, where R is a symmetric and antireflexive binary relation on G is called a _graph_. Elements of G are called _vertices_ of G. A graph G is _complete_ if any two vertices of G are related (we will write $(x,y) \in R$, $R(x,y)$, or xRy interchangeably). A sequence $\langle x_1, x_2, \cdots, x_n\rangle$ of vertices in G is called a _path (of length n-1)_ if $x_1Rx_2Rx_3R \cdots Rx_{n-1}Rx_n$. A path of length 1 is called an _edge_. A sequence $\langle x_1, x_2, \cdots x_n\rangle$ of vertices is called a _cycle (of length n)_ if $x_1Rx_2R \cdots \cdots Rx_{n-1}Rx_nRx_1$. If a graph G possesses cycles [of odd length] at all, we say that G has _girth_ [_odd girth_] g if the shortest cycles [of odd length] have length g. The _chromatic number_ of a graph G, _ch(G)_, is the smallest cardinal $\mathfrak{m}$ such that there is a homomorphism of G onto the complete graph with $\mathfrak{m}$ vertices, $\underline{C(\mathfrak{m})}$. An easy folklore result is

10.1. Proposition ("Ore's Theorem"). A graph G possesses no cycles of odd length iff $ch(G) = 2$.

The question of the independence of chromatic number and girth was settled conclusively by P. Erdös in [12], where it is shown using probabilistic methods that for any $k \geq 3$ and any $g \geq 3$ there exists a (finite) graph G of girth g and with $ch(G) = k$. Quite recently this result has been sharpened by B. Bollabás & N. Sauer [5]; they show that G exists with the above properties where in addition G is _uniquely_ _k-colourable_, i.e., there is precisely one partition of G which is the kernel of all homomorphisms onto $C(k)$. Again their arguments are probabilistic (and anything but constructive). W. Taylor took a different tack in [45]; he defines a graph E^n_ε on the Euclidean unit ball E^n by relating two points if their distance exceeds $2 - \varepsilon$. By taking n large enough and $\varepsilon > 0$ small enough (for pregiven k and g) E^n contains a finite subgraph $G(k,g)$ with chromatic number k and _odd_ girth at least g. The proof is not constructive in the strict sense, and the examples are on the surface weaker than those of Erdös; nevertheless an analysis of further properties of the $G(k,g)$'s would seem, due to the measure theoretic setting, to be more accessible than any further enlightening of the Erdös examples. Be this as it may, let G_n denote, for each $n \in \mathbb{N}$, $n \geq 3$, a fixed finite graph of chromatic number n and odd girth at least n. The following result, due to W. Taylor, yielded the first negation of the Mycielski Question in general. All succeeding constructions of counterexamples have been based upon this graph.

10.2. Proposition (W. Taylor [45]). Let $G = \langle G; R\rangle$ be the graph

$$G = G_3 \dot\cup G_4 \dot\cup G_5 \dot\cup \cdots \dot\cup G_n \dot\cup \cdots$$

where R is the union of all the relations on G_n, $n \geq 3$. Then G is atomic compact but not retract of any compact algebraic system.

For the rest of this chapter $G = \langle G; R\rangle$ will denote Taylor's graph defined above. Note that $ch(G)$ is infinite as $ch(G_n)$ cannot exceed $ch(G)$ for any n. Thus one half of Proposition 2 follows immediately from

10.3. Proposition (W. Taylor [45]). A graph which is a retract of some compact algebraic system has finite chromatic number.

A proof of the other half (that G is atomic compact) different from Taylor's proof in [45] can be found in the next section.

Using G one can quite easily construct an equationally compact algebra $\mathcal{B} = \langle B; f_1, f_2 \rangle$ which is not a retract of any compact algebra : The carrier set B is taken to be $G \dot{\cup} R \dot{\cup} \{e_1, e_2\}$, and f_i is the unary operation

$$f_i(b) = \begin{cases} g_i , & \text{if } b = (g_1, g_2) \in R \\ e_i , & \text{otherwise} \end{cases} , \quad i = 1,2 .$$

This example is also due to W. Taylor (private communication) and points out too the sense in which G.H. Wenzel's positive answer to the Mycielski Question for mono-unary algebras is best possible.

The next breakthrough was a semigroup counterexample (W. Taylor [48]). The idea is to construct a semigroup possessing G as a subreduct, and goes as follows :

Let S be the semigroup S_o/θ , where S_o is the semigroup freely generated by G in the equational class $\mathbb{K}$ of (associative) semigroups satisfying the identity $x_1x_2x_3 = x_4x_5x_6$, and θ is the congruence collapsing a and b if *each* of a and b is a product of two elements in G related by R . One can view S as the set

$$G \,\dot{\cup}\, G^2 \setminus R \,\dot{\cup}\, \{0\} \,\dot{\cup}\, \{\alpha\} \ ,$$

with multiplication defined by

$$a \cdot b = \begin{cases} \alpha & \text{if } a,b \in G \text{ and } (a,b) \in R \\ (a,b) & \text{of } a,b \in G \text{ and } (a,b) \notin R \\ 0 & \text{otherwise} \ . \end{cases}$$

The idea of the proof that S is equationally compact, i.e., that S is a retract of every ultrapower of S (cf. Proposition 2.3), is as follows :

For an ultrapower $T = S_F^I$ there exists a graph retraction of G_F^I onto G as G is atomic compact, which then extends to a semigroup epimorphism $f : T_o \longrightarrow S_o$, where T_o is the semigroup in $\mathbb{K}$ freely generated by G_F^I . The trick is then to retrieve T canonically as a homomorphic image of T_o , as then

$$\begin{array}{ccc} T_o & \xrightarrow{\ f\ } & S_o \\ \downarrow & & \downarrow \\ T = S_F^I & \dashrightarrow & S \end{array}$$

completes commutatively, yielding the desired retraction. Essential to the argument is the following fact:

(FP) There is a *finite* set of polynomials which when applied to the generating set G yield all of S ,

for then, and *only then*, is T also generated by G_F^I , i.e., *only then* will the canonical homomorphism $T_o \longrightarrow T$ be surjective,

In other words, the success of the proof depends not so much on any particularly semigroup theoretical aspects but rather on the fact that we are in an equational class in which the free algebras satisfy the property (FP) .

In a fashion totally analogous to W. Taylor's construction of S M.L. Kleǐner defines in [32] a ring A which he claims is also a counterexample. Rather disturbing is that he gives no indication of a proof, moreover A does not satisfy (FP) (so Taylor's proof does not work), and all direct requests on our part for a copy of the proof were put off.

In the next section we subject A to a closer scrutiny.

§11 A further counterexample

11.1. Definition (M.L. Kleǐner [32]). A is the ring $A_o/\mathfrak{A}$, where A_o is the ring freely generated by G in the equational class of commutative rings satisfying the identities $x + x = 0$ and $xyz = 0$, and $\mathfrak{A}$ is the ideal

$$\left\{ \sum_{i=1}^{m} g_i \cdot h_i \ ; \ g_i, h_i \in G \ , \quad g_i R h_i \ , \quad m \in 2\mathbb{N} \right\} .$$

Observe that one can view A^+ as the linear space over Z_2 with Basis

$$B = G \,\dot{\cup}\, \{\{g,h\} ; \ (g,h) \in G^2 \setminus R\} \,\dot{\cup}\, \{\alpha\}$$

and with multiplication in A defined by distributivity and the rule

$$a \cdot b = \begin{cases} \alpha & \text{if } a,b \in G \ , \ (a,b) \in R \\ \{a,b\} & \text{if } a,b \in G \ , \ (a,b) \notin R \\ 0 & \text{otherwise} \end{cases}$$

for $a,b \in B$. For brevity we will denote by M_+ the subgroup of A^+ generated by the subset M of A; also G^m will denote $G_3 \dot{\cup} G_4 \dot{\cup} \cdots \dot{\cup} G_m$ for $m \geq 3$. Note that $A^2 = \mathrm{Ann}(A)$ and that $A^+ = G_+ \oplus A^2{}_+$. For arbitrary $c \in A$ denote by $\underline{c}$ the set of basis vectors occurring with scalar coefficient 1 in the representation of c as a linear combination of B. Define now the binary relation

$$R_\alpha(a,b) \quad :\text{iff} \quad a \cdot b = \alpha \ , \qquad a,b \in A \ ,$$

and let

$$\mathfrak{A} = \langle A; +, R_\alpha \rangle .$$

Now the restriction of R to any subgroup H of A^+ turns

H into a substructure of $\mathfrak{A}$, however we will not notationally distinguish $\langle H; +\rangle$ from $\langle H; +, R_\alpha\rangle$ as the interpretation will always be clear from the context.

Our goal is to show that $\mathfrak{A}$ is atomic compact; we will go about this by defining a graph structure on the variable set $X = \mathrm{Var}(\Sigma)$ of a finitely solvable system Σ of atomic formulas over $\mathfrak{A}$ such that X has chromatic number 2 and such that any homomorphism of X onto an edge of $G \subseteq A$ is a solution of Σ in $\mathfrak{A}$. To do this successfully we must be allowed to concentrate on very special systems Σ; permission for this will be achieved via the following reduction steps. The reader is advised to read through Lemmas 2 to 9, skipping over the proofs first, and convince himself that they do imply Proposition 10.

First we must appeal to the Mycielski-Ryll-Nardzewski Theorem (2.4) which says (somewhat paraphrased) that it suffices to solve finitely solvable systems of atomic formulas over $\mathfrak{A}$ of the following restrictive type ("M.R.N. systems"):

$$\Sigma = \dot{\bigcup}(\Sigma_i \ ; i \in I) ,$$

where for each $i \in I$ Σ_i is a finite set of atomic formulas, and for each $i \neq j$ $\mathrm{Var}(\Sigma_i) \cap \mathrm{Var}(\Sigma_j) \subseteq \{x_o\}$.

Now atomic formulas over $\mathfrak{A}$ can be classified as the following types of ring equations over A:

Type I : $x \cdot y = \alpha$, i.e., $R_\alpha(x,y)$
Type I' : $x \cdot b = \alpha$, i.e., $R_\alpha(x,b)$
Type II : $x_1 + \cdots + x_n = c$,

where $x,y,x_1,\cdots,x_n$ are variables, $b,c \in A$ and $n \in \mathbb{N}$.

<u>11.2. Lemma</u>. If $\Sigma = \dot{\bigcup}(\Sigma_i;\ i \in I)$ is a finitely solvable M.R.N. system over $\mathfrak{A}$, then there is a finitely solvable system Σ_o of type I, I', II formulas over $\mathfrak{A}$ such that

(1) every variable occurring in some type II equation occurs in some type I or type I' equation,

and the solvability of Σ_o in $\mathfrak{A}$ implies the solvability of Σ in $\mathfrak{A}$.

<u>proof</u>: Fix i and suppose $x_1, \cdots, x_n$ are the variables occurring in Σ_i excluding x_o, and suppose $x_1, \cdots, x_t$ are those occurring in some type II formula but in no type I or I' formula (if there are none of this kind let Σ_i be). Suppose x_1 occurs in the formula $\phi_1^{(1)} \equiv x_1+y_1+ \cdots +y_k = c$. Obtain Σ_{i1} by replacing every occurrence of x_1 in $\Sigma_i \setminus \{\phi_1^{(1)}\}$ by $y_1+ \cdots + y_k + c$. Then the formula $\bigwedge\Sigma_i$ is equivalent to the formula $(\bigwedge\Sigma_{i1}) \wedge \phi_1^{(1)}$. It follows that $\Sigma' = (\Sigma \setminus \Sigma_i) \cup \Sigma_{i1}$ is a finitely solvable system in which x_1 does not occur, and any solution of Σ' yields a solution of Σ by plugging into $\phi_1^{(1)}$ the values taken on by $y_1, \cdots ,y_k$ to obtain an appropriate value for the remaining variable x_1. Now consider x_2. If x_2 occurs <u>only</u> in $\phi_1^{(1)}$ replace x_2 by an arbitrary constant, yielding the formula $\phi_1^{(2)}$, and all the above properties of Σ', defined now for $\phi_1^{(2)}$ instead of $\phi_1^{(1)}$, still hold. If, on the other hand, x_2 occurs in Σ_{i1}, say in

$$\phi_2^{(2)} \equiv x_2+z_1+ \cdots +z_l = d\ , \qquad \{z_1,\cdots,z_l\} \subseteq \{x_3,\cdots,x_n,x_o\},$$

replace every occurrence of x_2 in $\phi_1^{(1)}$ and $\Sigma_{i1} \setminus \{\phi_2^{(2)}\}$ by $z_1+ \cdots +z_l + d$ obtaining $\phi_1^{(2)}$ (from $\phi_1^{(1)}$) and Σ_{i2}

(from $\Sigma_{i1} \setminus \{\phi_2^{(2)}\}$) . Again

$$\Sigma'' = (\Sigma \setminus \Sigma_i) \cup \Sigma_{i2}$$

is a finitely solvable system in which x_1, x_2 do not occur, and any solution of which yields a solution of Σ by plugging into first $\phi_2^{(2)}$ and then into $\phi_1^{(2)}$ to obtain appropriate values for first x_2 and then x_1 . Proceeding inductively in this way we obtain formulas $\phi_1^{(t)}, \phi_2^{(t)}, \cdots, \phi_t^{(t)}$ and a set Σ_i' in which only $x_o, x_{t+1}, x_{t+2}, \cdots, x_n$ occur, such that

$$\Sigma' := (\Sigma \setminus \Sigma_i) \cup \Sigma_i'$$

is finitely solvable and such that any solution of Σ' yields values for $x_t, x_{t-1}, \cdots, x_1$ (by plugging recursively into $\phi_t^{(t)}, \phi_{t-1}^{(t)}, \cdots, \phi_1^{(t)}$) giving a solution to Σ.

If we now well-order I a trivial transfinite induction argument yields that we can apply this elimination process to each Σ_i successively; thus there exists a system Σ_o', finitely solvable in $\mathfrak{A}$, the solvability of which implies the solvability of Σ and is such that either $\Sigma_o := \Sigma_o'$ does the job or else <u>precisely</u> the variable x_o occurs in a type II formula without occurring in any type I or I' formula. If the latter holds, say x_o occurs in

$$\phi \equiv x_o + w_1 + \cdots + w_m = b$$

in Σ_o' ; replace now <u>every</u> occurrence of x_o in $\Sigma_o' \setminus \{\phi\}$ by $w_1 + \cdots + w_m + b$ obtaining a finitely solvable system Σ_o whose solvability in $\mathfrak{A}$ implies the solvability of Σ_o' (and hence that of Σ) in $\mathfrak{A}$ and moreover enjoying the property (1). ●

<u>11.3. Lemma</u>. If Σ is a finitely solvable system of type I, I', II formulas satisfying (1), then there is a system Σ_o of type I,I',II formulas satisfying (1) and

(2) Σ_o has constants from G_+ and is finitely solvable in G_+,

and is such that if Σ_o is solvable in G_+ then Σ is solvable in $\mathfrak{A}$.

<u>proof</u>: As $A^+ = G_+ \oplus A^2_+$ we can write, for any $c \in A$, $c = c_{G_+} + c_{A^2_+}$ uniquely with $c_{G_+} \in G_+$ and $c_{A^2_+} \in A^2_+$.
Now any type I' formula $R_\alpha(x,b)$ is equivalent in $\mathfrak{A}$ to $R_\alpha(x,b_{G_+})$, as $b_{A^2_+} \in \mathrm{Ann}(A)$; hence w.l.o.g. we may assume all type I' formulas in Σ already have constants in G_+.
Now set

$$\Sigma_o := \{R_\alpha(x^{(G_+)}, y^{(G_+)});\ R_\alpha(x,y) \text{ occurs in } \Sigma\}$$

$$\cup \{R_\alpha(x^{(G_+)}, b)\ ;\ R_\alpha(x,b) \text{ occurs in } \Sigma\}$$

$$\cup \{x_1^{(G_+)} + \dots + x_n^{(G_+)} = c_{G_+};\ x_1 + \dots + x_n = c \text{ occurs in } \Sigma\},$$

and set

$$\Sigma_1 := \{x_1^{(A^2_+)} + \dots + x_n^{(A^2_+)} = c\ ;\ x_1 + \dots x_1 = c \in \Sigma\}.$$

Obviously Σ_o satisfies (1) if Σ does, and the finite solvability of Σ in $\mathfrak{A}$ implies that Σ_o is finitely solvable in G_+ and Σ_1 is finitely solvable in A^2_+ (if s is the substitute for the variable x in some solution, let $x^{(G_+)}$ (resp. $x^{(A^2_+)}$) take on the value s_{G_+} (resp. $s_{A^2_+}$) in the corresponding formulas in Σ_o (resp. Σ_1)).
Thus Σ_o satisfies (2) also. But as Σ_1 is a finitely solvable system of linear equations over the (equationally compact!) vector space A^2_+, Σ_1 is solvable in A^2_+.

Thus if Σ_o is solvable in G_+, any solution of Σ_1 in $A^2{}_+$ yields (by reversing the process described in the last parenthetical comment) a solution of Σ in $\mathfrak{A}$. •

<u>11.4. Lemma</u>. Let Σ be a system of type I, I', II formulas satisfying (1) and (2). Then there is a system Σ_o of type I, I', II formulas satisfying (1), (2), and

(3) for all variables occurring in Σ and all $c \in G_+$, the system $\Sigma_o \cup \{x = c\}$ is not finitely solvable in G_+,

and is such that if Σ_o is solvable in G_+, then so is Σ.

<u>proof</u>: Define

$$\mathcal{F} = \{\Sigma' ;\ \Sigma' \text{ is finitely solvable in } G_+ \text{ and is obtained from } \Sigma \text{ by replacing, for some subset } \{x_i;\ i \in I\} \subseteq \mathrm{Var}(\Sigma)\text{, \underline{each} occurrence of } x_i \text{ by some } c_i \in G_+,\ i \in I\}.$$

Partially order $\mathcal{F}$ by

$\Sigma' \leq \Sigma''$:iff if in Σ' the variable x was replaced by $c \in G_+$, then the same is true for Σ''.

$\mathcal{F}$ is not empty, as $\Sigma \in \mathcal{F}$, and $\mathcal{F}$ is clearly inductive, as finite solvability requires looking at only finitely many equations, and hence only finitely many variable substitutions at a time. Thus by Zorn's Lemma $\mathcal{F}$ possesses a maximal element Σ_o. Now <u>every</u> member of $\mathcal{F}$ (after throwing out all identities possibly present) satisfies all the requirements stated in the Lemma except possibly (3). But Σ_o satisfies (3) also by maximality. •

<u>11.5. Definition</u>. Let $\mathcal{B}$ be an algebraic system and let Σ be a system of first order formulas over $\mathcal{B}$. For a variable $x \in Var(\Sigma)$ denote by $\underline{Sol_{\mathcal{B}}(\Sigma; x)}$ those substitutes for x in B occurring in elements of $Sol_{\mathcal{B}}(\Sigma)$. A variable $x \in Var(\Sigma)$ is <u>finitely bounded by Σ in $\mathcal{B}$</u> if there is a finite subset Σ' of Σ such that $Sol_{\mathcal{B}}(\Sigma; x)$ is finite.

<u>11.6. Lemma</u>. Let Σ be a system enjoying the properties holding for Σ_o in Lemma 4. Then Σ satisfies

(4) no variable in $Var(\Sigma)$ is finitely bounded by Σ in G_+,

and

(5) Σ consists only of type I and type II formulas .

<u>proof</u>: (4) : Suppose $x \in Var(\Sigma)$ is finitely bounded by Σ in G_+, say

$$Sol_{G_+}(\Sigma'; x) = \{c_1, \cdots, c_n\} \subseteq G_+$$

with $\Sigma' \subseteq \Sigma$ finite. By (3) there exists, for each $i = 1, \cdots, n$, a finite subset Σ_i of Σ such that

$$\Sigma_i \cup \{x = c_i\} \text{ is not solvable in } G_+ .$$

But then $\Sigma' \cup \Sigma_1 \cup \cdots \cup \Sigma_n$ is not solvable in G_+, contradicting the finite solvability of Σ.

(5) : This will follow from (4) if we show that the variable x is finitely bounded by the type I' formula $R_\alpha(x,b)$ in G_+. Indeed, one easily checks that from $a \cdot b = \alpha$, $a \in G_+$, it follows that $\underline{a} \cup \underline{b} \subseteq G_i$ for some $i \in \mathbb{N}$. Thus

$$|Sol_{G_+}(R_\alpha(x,b); x)| \leq |(G_i)_+| < \aleph_o . \qquad \bullet$$

11.7. Definition. A partition p of the type II formula $x_1 + \cdots + x_n = c \equiv \phi$ is a set of type II formulas

$$p = \{x_{11} + \cdots + x_{1n_1} = c_1 \ , \ x_{21} + \cdots + x_{2n_2} = c_2 \ , \ \cdots \ , \ x_{m1} + \cdots + x_{mn_m} = c_m\}$$

where $\{x_1, \cdots, x_n\} = \dot{\bigcup}(\{x_{ij}\} ; 1 \le j \le n_i, 1 \le i \le m)$ and $\underline{c} = \underline{c}_1 \dot{\cup} \cdots \dot{\cup} \underline{c}_m$. Denote by $P(\phi)$ the set of partitions of ϕ and partially order $P(\phi)$ by $p_1 \le p_2$:iff every member of p_1 is partitioned by members of p_2 . For $m = 1$ in the above p is trivial, and for $m = 2$ p is simple.

11.8. Lemma. Let Σ be a system enjoying the properties holding for Σ_o in Lemma 4. Then there is a system Σ_o satisfying (1) - (5) and

(6) For any non-trivial partition p of any type II formula in Σ_o , $\Sigma_o \cup p$ is not finitely solvable in G_+ ,

and is such that the solvability of Σ_o in G_+ implies the same for Σ .

proof: Define

$\mathcal{F}$:= { Σ' ; Σ' is finitely solvable in G_+ and is obtained from Σ by replacing each type II $\phi \in \Sigma$ by a (possibly trivial) partition, say $p_{\phi,\Sigma'}$, of ϕ } .

Partially order $\mathcal{F}$ by

$$\Sigma' \le \Sigma'' \quad \text{:iff} \quad p_{\phi,\Sigma'} \le p_{\phi,\Sigma''} \text{ for each type II } \phi \in \Sigma.$$

Again it is easy to see that $\mathcal{F}$ is not empty and inductively ordered, hence Zorn's Lemma gives a maximal element, say Σ_o . Clearly Σ_o satisfies (1), (2), and (6). Now obviously any solution of a partition of ϕ is a solution of ϕ , and hence (3) follows for Σ_o from the fact that (3) held for Σ, namely if $\{\phi_1, \cdots, \phi_m\} \cup \{x = c\}$ is a finite subset of $\Sigma \cup \{x = c\}$ which is not solvable, then

$$\bigcup(p_{\phi_i,\Sigma_o} ; i = 1, \cdots, m) \cup \{x = c\}$$

is a finite non-solvable subset of $\Sigma_o \cup \{x = c\}$. By the above observation it follows, too, that any solution of Σ_o is a solution of Σ, and so, with the help of Lemma 4, the claim is proved. •

<u>11.9. Lemma</u>. A system Σ of formulas satisfying (1) - (6) satisfies

(7) Every type II formula in Σ is of the form $x_1 + \cdots + x_n = 0$ with n even.

<u>proof</u>: We have already noted in the proof of Lemma 6 that for any elements $a,b \in G_+$ with $a \cdot b = \alpha$, $\underline{a} \cup \underline{b} \subseteq G_i$ for some i . A moment's reflection yields the following sharpening of this fact:

(*) for $a,b \in G_+$, $a \cdot b = \alpha$ iff $|\underline{a}|$ and $|\underline{b}|$ are odd and for every $a_o \in \underline{a}$ and $b_o \in \underline{b}$, $R(a_o,b_o)$.

Now suppose the formula $\phi \equiv x_1 + \cdots + x_n = c$ occurs in Σ . ($n \geq 2$ by (3)!) By (6) there exists for each simple partition $p \in P(\phi)$ a finite subset Σ_p of Σ such that $\Sigma_p \cup p$ is not solvable in G_+ . By (1) and (5) there is a type I formula $\phi_i \in \Sigma$ with $x_i \in \mathrm{Var}(\phi_i)$, $i = 1, \cdots, n$. Set

$$\Sigma_o := \bigcup(\Sigma_p;\ p \in P(\phi),\ p \text{ simple}) \cup \{\phi_1, \cdots, \phi_n, \phi\} .$$

Now by (2) $c \in G^m{}_+$ for some $m \in \mathbb{N}$. Assume that $c \neq 0$. Let $s_1, \cdots, s_n$ be the values taken on by $x_1, \cdots, x_n$ in an arbitrary but fixed solution of Σ_o in G_+ . We claim that $s_1, \cdots, s_n \in G^m{}_+$. If we show this then it follows that $x_1, \cdots, x_n$ are all finitely bounded by Σ in G_+ , as $G^m{}_+$ is finite, contradicting (4), and thus $c = 0$ as desired. Now $s_1 + \cdots + s_n = c \in G^m{}_+$; thus for at least one s_{i_1} , some element of $\underline{s}_{i_1}$ lies in some G_i with $i \leq m$, and if $s_{i_1}, \cdots s_{i_k}$ are those s_i's satisfying this, it follows that $\underline{s}_{i_1} \cup \cdots \cup \underline{s}_{i_k} \subseteq G^m$ by ($*$) and the fact that each s_i satisfies ϕ_i . Therefore if our claim were not true there would be s_i's remaining, say $s_{j_1}, \cdots, s_{j_l}$ such that $(\underline{s}_{j_1} \cup \cdots \cup \underline{s}_{j_l}) \cap G^m = \emptyset$ and it follows that

$$s_{i_1} + \cdots + s_{i_k} = c \quad \text{and} \quad s_{j_1} + \cdots + s_{j_l} = 0$$

which means that our solution of Σ_o is a solution of $\Sigma_p \cup p$ where p is the simple partition

$$\{x_{i_1} + \cdots + x_{i_k} = c\ ,\quad x_{j_i} + \cdots + x_{j_l} = 0\} ,$$

a contradiction. Thus $s_1 + \cdots + s_n = 0$, and by ($*$) each $|\underline{s}_i|$ is odd; if n were odd, this would imply that zero can be represented as an odd sum of basis vectors in a linear space over Z_2 , which is absurd. Thus n is even. $\bullet$

The Mycielski-Ryll-Nardzewski Theorem (2.4) together with Lemmas 2 through 9 yield

<u>11.10. Proposition</u>. $\mathfrak{A}$ is atomic compact if every system of formulas satisfying (1) - (7) is solvable in G_+ .

Our goal, then, is to show that the hypothesis of Proposition 10 is true. If we do this we will have given, by the way, an independent (albeit lengthy!) proof that W. Taylor's graph G is atomic compact. Indeed, Lemmas 2 through 9 say that for any finitely solvable M.R.N. system over $\mathfrak{A}$ - and in particular for any M.R.N. system Σ_G over G , finitely solvable in G - there is a system $\tilde{\Sigma}_G$ satisfying (1) - (7) whose solvability in G_+ <u>implies</u> the solvability of Σ_G in $\mathfrak{A}$. But remembering ($*$) , any solution of Σ_G in $\mathfrak{A}$ yields easily a solution of Σ_G in G .

So let Σ be a system satisfying (1) - (7), and let $X = \mathrm{Var}(\Sigma)$. Define a graph structure on X as follows: First, denote by $x \sim_\phi y$ the situation that $\phi \in \Sigma$, $x \neq y$, and $x,y \in \mathrm{Var}(\phi)$; then set

$$\varepsilon(\phi,x,y) = \begin{cases} 1 & \text{if } x \sim_\phi y \text{ and } \phi \text{ is type I} \\ 0 & \text{otherwise} \end{cases}$$

for all $x,y \in X$ and $\phi \in \Sigma$, and define

$$(x,y) \in R_X \quad :\text{iff} \quad \exists_{\substack{m \in \mathbb{N} \\ x=z_0,z_1,\cdots,z_m=y \in X \\ \phi_1,\cdots,\phi_m \in \Sigma}} \begin{cases} z_{i-1} \sim_{\phi_i} z_i \ , \ i=1,\cdots,m \\ \qquad \text{and} \\ \sum_{i=1}^{m} \varepsilon(\phi_i,z_{i-1},z_i) \ \text{ is odd.} \end{cases}$$

Intuitively, xR_Xy if we can find a path of variables leading from x to y , meaning that each adjacent pair occurs in a common $\phi \in \Sigma$, with type I equations being used as "stepping stones" an odd number of times.

<u>11.11. Lemma</u>. $\langle X; R_X \rangle$ is a graph with no cycles of odd length.

<u>proof</u>: R_X is trivially symmetric, as the relations $\sim_\phi$ are symmetric. Antireflexivity and the remaining claim will be proved if we show that there is no tuple $(x_1, \cdots, x_m) \in X^m$ with m odd satisfying

$$x_1 \; R_X \; x_2 \; R_X \; \cdots \; R_X \; x_{m-1} \; R_X \; x_m \; R_X \; x_1 \; .$$

(Since if $x \; R_X \; x$ then (x,x,x) would be such a tuple; thus $\langle X; R_X \rangle$ will be a graph and of course any cycle of odd length in $\langle X; R_X \rangle$ would then yield such a tuple too.)

So suppose there were such a tuple $(x_1, \cdots, x_m)$. Fix i $(1 \leq i \leq m)$. From $x_i \; R_X \; x_{i+1}$ (let $x_{m+1} := x_1$) we have by definition

$$x_i = z_o \sim_{\phi_1} z_1 \sim_{\phi_2} z_2 \sim_{\phi_3} \cdots \sim_{\phi_{m_i}} z_{m_i} = x_{i+1} \; .$$

For any $j = 1, \cdots, m_i$ we have one of the following cases:

(a) $z_{j-1} \sim_{\phi_j} z_j$ with ϕ_j type II .

(b) $z_{j-1} \sim_{\phi_j} z_j$ with ϕ_j type I (i.e., $\phi_j \equiv z_{j-1} \cdot z_j = \alpha$)

If (a) holds then (1), (5), and (6) guarantee the existence of a finite subset Σ_{ij} of Σ containing for each w in $Var(\phi_j)$ a type I equation in which w occurs, and is such that <u>no</u> solution of Σ_{ij} satisfies <u>any</u> non-trivial partition of ϕ_j .

If (b) holds set $\Sigma_{ij} := \{\phi_j\}$. For both (a) and (b) let $k_j := |Var(\phi_j)|$.

Now set $\Sigma_i = \bigcup(\Sigma_{ij} \; ; \; j = 1, \cdots, m_i)$ and consider an arbitrary but fixed solution $s: X \to G_+$ of Σ_i in G_+ .

Consider for an arbitrary j again the case (a) : Let μ be the reflexive symmetric binary relation on $Var(\phi_j)$ defined by

$$w \mu w' \ :\text{iff} \ \ \underline{s(w)} \cap \underline{s(w')} \neq \emptyset ,$$

and let $\bar{\mu}$ be the transitive closure of μ. Observe that the equivalence relation $\bar{\mu}$ is $Var(\phi_j) \times Var(\phi_j)$ since if $\bar{\mu}$ properly partitioned $Var(\phi_j)$, s would be a solution of the corresponding non-trivial partition of ϕ_j, contradicting the choice of Σ_{ij}. Now for $w \mu w'$ there exists a path in G (see figure) of even length ≤ 4 connecting any element of $\underline{s(w)}$ with any element of $\underline{s(w')}$. As $z_{j-1} \bar{\mu} z_j$ ($\bar{\mu}$ relates all variables!) we conclude there is a path of even length $\leq 4 \cdot k_j$ connecting any element of $\underline{s(z_{j-1})}$ with any element of $\underline{s(z_j)}$.

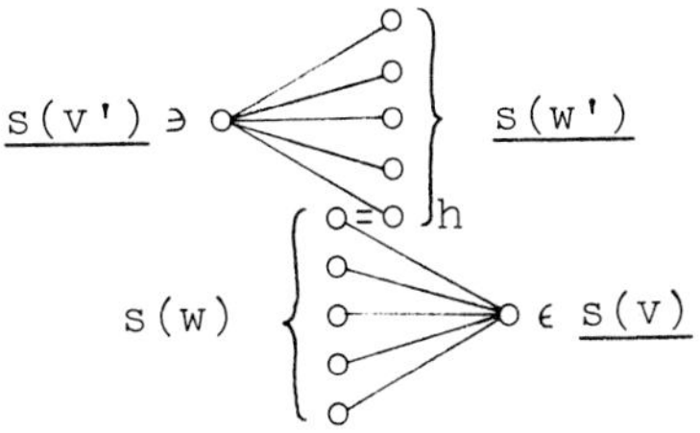

$h \in \underline{s(w)} \cap \underline{s(w')}$,

$R_\alpha(w,v)$, $R_\alpha(w',v') \in \Sigma_{ij}$

If case (b) holds then by (*) there is a path of length one connecting any element of $\underline{s(z_{j-1})}$ with any element of $\underline{s(z_j)}$. By the definition of R_X case (b) occurs for an odd number of j's , hence we conclude there is a path of odd length $\leq l_i := 4 \cdot (k_1 + \cdots + k_{m_i})$ connecting any element of $\underline{s(x_i)}$ with any element of $\underline{s(x_{i+1})}$. Setting now $l := l_1 + \cdots + l_m$ and $\Sigma_o = \Sigma_1 \cup \cdots \cup \Sigma_m$ it follows that for an arbitrary solution $s : X \longrightarrow G_+$ of the finite set $\Sigma_o \subseteq \Sigma$, <u>every</u> element of $\underline{s(x_1)}$ is a

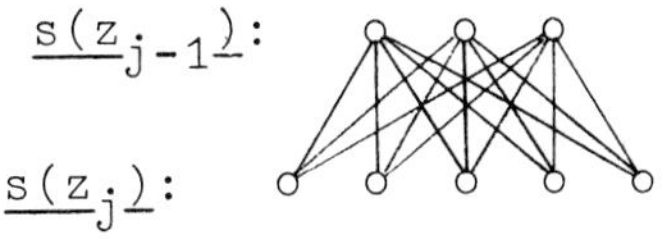

vertex in a cycle of odd length ≤ 1 , as m was odd, and thus $s(x_1) \in G^1{}_+$. We conclude that x_1 is finitely bounded by Σ in G_+ , a contradiction. ●

It follows now by Ore's Theorem (10.1) that $ch(\langle X; R_X \rangle)$ is 2 and thus there is a homomorphism

$$s : \langle X; R_X \rangle \longrightarrow \langle E; R_\alpha \rangle$$

where E is an arbitrary (fixed) edge of G . We claim that s is a solution of Σ in G . Indeed, for $R_\alpha(x,y)$ in Σ we have $x\ R_X\ y$ and hence $s(x)\ R_\alpha\ s(y)$, i.e., $(s(x),s(y))$ is a solution of $R_\alpha(x,y)$. And if $x_1 + \cdots + x_n = 0 \equiv \phi \in \Sigma$ consider x_i, x_j for $i \neq j$. By (1) and (5) $\phi' \equiv R_\alpha(x_i,z) \in \Sigma$ for some $z \in X$. But then

$$x_i \sim_{\phi'} z \quad \text{and} \quad x_j \sim_\phi x_i \sim_{\phi'} z$$

and so $$x_i\ R_X\ z \quad \text{and} \quad x_j\ R_X\ z$$

whence $$s(x_i)\ R\ s(z) \text{ and } s(x_j)\ R\ s(z)$$

i.e., $$s(x_i) \neq s(z) \neq s(x_j)$$

and so $s(x_i) = s(x_j)$. As n is even by (7) we have

$$s(x_1) + \cdots + s(x_n) = n \cdot s(x_1) = 0 ,$$

and so s is a solution of ϕ . We have proven that $\mathfrak{A}$ is atomic compact.

Now suppose that $\mathfrak{A}$ is a retract of some compact algebraic system. Then the reduct $\langle A; R_\alpha \rangle$ would be too. But the graph $\langle A; R_\alpha \rangle$ has infinite chromatic number because its subgraph $\langle G; R \rangle$ has. This contradicts Taylor's Theorem (10.3). We have proven

11.12. Theorem. There exists a commutative ring A and an algebraic system $\mathfrak{A}$ such that

(1) $\mathfrak{A}$ is a reduct of A and the additive group A^{+} is a reduct of $\mathfrak{A}$,

(2) $\mathfrak{A}$ is atomic compact,

(3) $\mathfrak{A}$ is not a retract of any compact algebraic system.

The question left undecided is whether the ring A is itself equationally compact, thus yielding a "no" to the Mycielski Question in the class of commutative rings. That A is indeed equationally compact is precisely the claim made by Kleĭner in [32]. If A were equationally compact, then there would be a counterexample too in the class of commutative unital rings. Indeed, $Z_2 * A$ is unital and would be equationally compact by Proposition 3.10, and, as is easily checked, the equation $x \cdot y = \alpha$ again defines a graph structure on $Z_2 * A$ and so $Z_2 * A$ also possesses as a reduct a graph of infinite chromatic number.

It is possible to enrich the structure of $\mathfrak{A}$ at least somewhat while retaining atomic compactness. If, for $c \in A$, the relation R_c is defined by $(x,y) \in R_c$:iff $x \cdot y = c$, then

$$\tilde{\mathfrak{A}} := \langle A;\, +,\, \{R_c\}_{c \in A} \rangle$$

is still an atomic compact reduct of the ring A. Indeed, one checks that the equation $x \cdot y = 0$ is equivalent over $\tilde{G}_+ := \langle G_+;\, +,\, \{R_c\}_{c\in A}\rangle$ to the positive formula

$$x=0 \vee y=0 \vee (\exists x_1)(\exists x_2)(\exists y_1)(\exists y_2)\,\{\, x=x_1+x_2 \wedge y=y_1+y_2 \wedge$$
$$[(x\cdot y_1=\alpha \wedge x\cdot y_2=\alpha) \vee (x_1\cdot y=\alpha \wedge x_2\cdot y=\alpha) \vee$$
$$(x_1\cdot y_1=\alpha \wedge x_1\cdot y_2=\alpha \wedge x_2\cdot y_1=\alpha \wedge x_2\cdot y_2=\alpha)]\,\},$$

and for arbitrary c one easily finds a positive formula over $\langle G_+; +, R_\alpha, R_0 \rangle$ equivalent to $x \cdot y = c$ over $\widetilde{G}_+$. Thus Weglorz' Theorem (2.3) and a reduction à la Lemma 3 from $\widetilde{\mathfrak{A}}$ to $\widetilde{G}_+$ yield the claim.

There are no known constructive examples of graphs possessing the properties of G we required. Thus there may well be no way of deciding the validity of Kleǐner's claim one way or the other. If, however, a verification can be found that A is equationally compact, then our Theorem may well provide a useful reduction step in the proof. For this reason we kept the discussion as broad as possible; either Erdös' probabilistic and highly unconstructive examples (or Bollabâs' & Sauer's refinement thereof), where the girth conditions are stricter than actually needed so far, may be taken to define the underlying graph G , or Taylor's examples may be used, and which have the attraction that the relations are defined concretely by the Euclidean distance function on the n-dimensional unit ball. Perhaps a deeper graph theoretic analysis within one or the other of these two classes of examples may provide enough new information to settle the question : Is the ring A , with perhaps G suitably restricted, equationally compact ?

RECURRING NOTATIONS

$\mathbb{N}$	set of natural numbers
$\mathbb{N}_o$	$\mathbb{N} \cup \{0\}$
$\mathbb{Z}$	ring of integers
$\mathbb{Q}$	field of rational numbers
$\mathbb{R}$	field of real numbers
$\mathbb{P}$	set of positive prime integers
ω_o	first infinite ordinal
$\mathbb{Z}(p^{\infty})$	p-th Prüfer group (multiplicative group of p^n-th roots of unity, $n \in \mathbb{N}$)
$\mathbb{Z}_p^*$	ring of p-adic integers
Z_n	ring of integers modulo n
$J(R)$	Jacobson radical of the ring R
R^+	additive group underlying the ring R
R_1	unital ring corresponding to the ring R with identity
$GF(q)$	Galois field with q elements
$\twoheadrightarrow$	epimorphism
$\rightarrowtail$	monomorphism
$\rightarrowtail\!\!\!\!\twoheadrightarrow$, $\cong$	isomorphism, isomorphic
$\dot{\cup}$	disjoint union
$[X]$	subalgebra generated by X
$D(R)$	maximal divisible subgroup of R^+
$Ann(R)$	annihilator of R
R_n	ring of $n \times n$ matrices over R $(n \geq 2)$
$Z(R)$	center of the ring R
$Ht(R)$	heart of the subdirectly irreducible ring R
	Classes of algebras and class operators are in bold face:
$\mathbb{H}(\mathbb{K})$	homomorphic images of members of $\mathbb{K}$
$\mathbb{S}(\mathbb{K})$	substructures of members of $\mathbb{K}$
$\mathbb{P}(\mathbb{K})$	direct products of members of $\mathbb{K}$

$\mathbb{P}_s(\mathbb{K})$	subdirect products of members of $\mathbb{K}$
$\mathbb{I}(\mathbb{K})$	isomorphic images of members of $\mathbb{K}$
$\mathbb{P}_u(\mathbb{K})$	ultraproducts of members of $\mathbb{K}$
$\mathbb{V}(\mathbb{K})$	equational class generated by $\mathbb{K}$ (= $\mathbb{HSP}(\mathbb{K})$)
$\mathbb{K}(\tau)$	class of algebraic structures of type τ
$P(\tau)$	polynomial symbols of type τ
$\mathcal{L}(\mathfrak{A})$	congruence lattice of the algebra $\mathfrak{A}$
$C_c(\mathfrak{A})$	set of closed congruences on the algebra $\mathfrak{A}$
$Id(\mathbb{V})$	set of polynomial identities satisfied in $\mathbb{V}$
Θ_F	congruence induced by the filter F
$\Theta(a,b)$	principal congruence generated by a,b
$\Theta\|_A$	congruence restricted to $\mathfrak{A}$
ω, ι	the smallest, resp. largest congruence
$P(\mathfrak{A})$	set of principal congruences on $\mathfrak{A}$
Spec $\mathfrak{A}$	set of maximal congruences on $\mathfrak{A}$
$E(r,s)$	set of maximal congruences identifying r,s
$D(r,s)$	set of maximal congruences distinguishing r,s
$\mathfrak{c}(\mathfrak{A})$	class of quasi-compactifications of $\mathfrak{A}$
$\chi(R)$	characteristic of the ring R
$ch(G)$	chromatic number of the graph G
$\underline{c}$	set of basis elements representing the vector c
$Var(\Sigma)$	set of variables occurring in members of Σ
X_+	subgroup generated by X
$\bullet$	end of proof

BIBLIOGRAPHY

[1] S. Balcerzyk, On the algebraically compact groups of I. Kaplansky, Fund. Math. 44 (1957) 91-93

[2] B. Banaschewski, Injectivity and essential extensions in equational classes of algebras, Queen's Papers on Pure and Applied Mathematics 25 (1970) 131-147

[3] ————, On equationally compact extensions of algebras, Alg. Univ. 4 (1974) 20-35

[4] ———— & E. Nelson, Equational compactness in equational classes of algebras, Alg. Univ. 2 (1972) 152-165

[5] B. Bollabás & N. Sauer, Uniquely colourable graphs with large girth, Canad. J. Math. 28 (1976) 1340-1344

[6] N. Bourbaki, Élements de mathématique, Livre III, Topologie générale, Chap.III Paris (1951)

[7] S. Bulman-Fleming, On equationally compact semilattices, Alg. Univ. 2 (1972) 146-151

[8] ———— & H. Werner, Equational compactness in quasi-primal varieties, Alg. Univ. 7 (1977) 33-46

[9] S. Burris, Boolean Powers, Alg. Univ. 5 (1975) 341-360

[10] A. Day, Injectivity in equational classes of algebras, Canad. J. Math. 24 (1972) 209-220

[11] J. Dugundji, Topology, Allyn and Bacon (1966)

[12] P. Erdös, Graph theory and probability, Canad. J. Math. 11 (1959) 34-38

[13] T. Frayne, A.C. Morel & D.S. Scott, Reduced direct products, Fund. Math. 51 (1962) 195-228

[13a] S. Gacsalyi, On pure subgroups and direct summands of abelian groups, Publ. Math. 4 (1955) 88-92

[14] I. Gelfand, Normierte Ringe, Rec. Math. (Math. Sbornik) 9 (51) (1941) 3-24

[15] M. Gould & G. Grätzer, Boolean extensions and normal subdirect powers of finite universal algebras, Math. Z. 99 (1967) 16-25

[16] G. Grätzer, Universal algebra, Van Nostrand, Princeton, N.J. (1968)

[17] D. C. Haines, Injective objects in the category of p-rings, Proc. AMS 42 (1974) 57-60

[18] D. K. Haley, On compact commutative noetherian rings, Math. Ann. 189 (1970) 272-274

[19] ———, Equationally compact artinian rings, Canad. J. Math. 25 (1973) 273-283

[20] ———, Equational compactness in rings with chain conditions (Dissertation), Manuskr. d. Fakultät für Mathematik u. Inf. 30 (1972)

[21] ———, A note on compactifying artinian rings, Canad. J. Math. 26 (1974) 580-582

[22] ———, Equational compactness and compact topologies in rings satisfying A.C.C., Pac. J. Math. 62 (1976) 99-115

[23] P. R. Halmos, Injective and projective Boolean algebras, in Lattice Theory, Proc. Symp. Pure Math. II Amer. Math. Soc. Providence (1961) 114-122

[24] I. N. Herstein, Noncommutative rings, The Carus Mathematical Monographs 15, John Wiley & Sons (1968)

[25] A. Hulanicki, On algebraically compact groups, Bull. Acad. Polon. Sci. Ser. Sci. Math. Astron. Phys. 10 (1962) 71-75

[26] N. Jacobson, Structure of rings, AMS Colloquium Publications 37 (1956)

[27] B. Jónsson, Algebras whose congruence lattices are distributive, Math. Scand. 21 (1967) 110-121

[28] I. Kaplansky, Topological rings, Amer. J. Math. 69 (1947) 153-183

[29] ———, Infinite abelian groups, University of Michigan Press, Ann Arbor (1954)

[30] K. Keimel & H. Werner, Stone duality for varieties generated by quasi-primal algebras, Memoirs AMS 148 (1974) 59-85

[31] J.L. Kelley, General topology, The University Series in Higher Mathematics, Van Nostrand (1955)

[32] M. L. Kleǐner, An example of an atomic-compact ring which is not a retract of any compact topological ring, Abstract from 3rd All-union Conference on Mathematical Logic, Novosibirsk (1974) (translation by W. Taylor)

[33] J. Loś, Quelques remarques théorêmes et problèmes sur les classes definissable d'algêbres, in Mathematical interpretation of formal systems, North-Holland, Amsterdam (1955) 98-113

[34] ———, Abelian groups that are direct summands of every abelian group which contains them as pure subgroups, Fund. Math. 44 (1957) 84-90

[35] N. McCoy, The theory of rings, Macmillan (1964)

[36] G. Michler & R. Wille, Die primitiven Klassen arithmetischer Ringe, Math. Z. 113 (1970) 369-372

[37] B. Müller, On Morita duality, Canad. J. Math. 21 (1969) 1338-1347

[38] J. Mycielski, Compactifications of general algebras, Colloq. Math. 13 (1964) 1-9

[39] ——————— & C. Ryll-Nardzewski, Equationally compact algebras II, Fund. Math. 61 (1968) 271-281

[40] A. Pixley, Functionally complete algebras generating distributive and permutable classes, Math. Z. 114 (1970) 361-372

[41] ———————, The ternary discriminator function in universal algebra, Math. Ann. 191 (1971) 167-180

[42] R. Raphael, Injective rings, Commun. in Alg. 1(5) (1974) 403-413

[43] P. Ribenboim, Rings and modules, Interscience Tracts in Pure and Applied Mathematics, John Wiley & Sons (1969)

[44] F. Szász, Über artinsche Ringe, Bull. Acad. Polon. Sci. Ser. math. astr. phys. 11 (1963) 351-354

[45] W. Taylor, Atomic compactness and graph theory, Fund. Math. 65 (1969) 139-145

[46] ———————, Some constructions of compact algebras, Ann. Math. Logic 3 (1971) 395-437

[47] ———————, Residually small varieties, Alg. Univ. 2 (1972) 33-53

[48] ———————, On equationally compact semigroups, Semigroup Forum 5 (1972) 81-88

[49] ———————, Pure compactifications in quasi-primal varieties, Canad. J. Math. 28 (1976) 50-62

[50] R. B. Warfield, Jr., Purity and algebraic compactness for modules, Pacific J. Math. 28 (1969) 699-719

[51] S. Warner, Compact rings and Stone-Čech compactifications, Arch. Math. 11 (1960) 327-332

[52] ————, Compact rings, Math. Ann. 145 (1962) 52-63

[53] B. Weglorz, Equationally compact algebras (I), Fund. Math. 59 (1966) 289-298

[54] ————, Equationally compact algebras (III), Fund. Math. 60 (1967) 89-93

[55] G. H. Wenzel, Relative solvability of polynomial equations in universal algebras, Habilitationsschrift an der Universität Mannheim (WH) (1970)

[56] ————, Subdirect irreducibility and equational compactness in unary algebras $\langle A;f\rangle$, Arch. Math. 21 (1970) 256-264

[57] ————, On $(\mathfrak{T},\mathfrak{A},\mathfrak{m})$-atomic compact relational systems, Math. Ann. 194 (1971) 12-18

[58] H. Werner, Eine Charakterisierung funktional vollständiger Algebren, Arch. Math. 21 (1970) 381-385

[59] ————, Algebraic representation and model-theoretic properties of algebras with the ternary discriminator (Habilitationsschrift), Preprint Nr. 237 TH Darmstadt Fachbereich Math. (1976)

[60] O. Zariski & P. Samuel, Commutative algebra Vol I & II, Van Nostrand, Princeton (1960)

[61] D. Zelinski, Linearly compact modules and rings, Amer. J. Math. 75 (1953) 79-90

SUBJECT INDEX

Vol. 580: C. Castaing and M. Valadier, Convex Analysis and Measurable Multifunctions. VIII, 278 pages. 1977.

Vol. 581: Séminaire de Probabilités XI, Université de Strasbourg. Proceedings 1975/1976. Edité par C. Dellacherie, P. A. Meyer et M. Weil. VI, 574 pages. 1977.

Vol. 582: J. M. G. Fell, Induced Representations and Banach *-Algebraic Bundles. IV, 349 pages. 1977.

Vol. 583: W. Hirsch, C. C. Pugh and M. Shub, Invariant Manifolds. IV, 149 pages. 1977.

Vol. 584: C. Brezinski, Accélération de la Convergence en Analyse Numérique. IV, 313 pages. 1977.

Vol. 585: T. A. Springer, Invariant Theory. VI, 112 pages. 1977.

Vol. 586: Séminaire d'Algèbre Paul Dubreil, Paris 1975–1976 (29ème Année). Edited by M. P. Malliavin. VI, 188 pages. 1977.

Vol. 587: Non-Commutative Harmonic Analysis. Proceedings 1976. Edited by J. Carmona and M. Vergne. IV, 240 pages. 1977.

Vol. 588: P. Molino, Théorie des G-Structures: Le Problème d'Equivalence. VI, 163 pages. 1977.

Vol. 589: Cohomologie l-adique et Fonctions L. Séminaire de Géométrie Algébrique du Bois-Marie 1965–66, SGA 5. Edité par L. Illusie. XII, 484 pages. 1977.

Vol. 590: H. Matsumoto, Analyse Harmonique dans les Systèmes de Tits Bornologiques de Type Affine. IV, 219 pages. 1977.

Vol. 591: G. A. Anderson, Surgery with Coefficients. VIII, 157 pages. 1977.

Vol. 592: D. Voigt, Induzierte Darstellungen in der Theorie der endlichen, algebraischen Gruppen. V, 413 Seiten. 1977.

Vol. 593: K. Barbey and H. König, Abstract Analytic Function Theory and Hardy Algebras. VIII, 260 pages. 1977.

Vol. 594: Singular Perturbations and Boundary Layer Theory, Lyon 1976. Edited by C. M. Brauner, B. Gay, and J. Mathieu. VIII, 539 pages. 1977.

Vol. 595: W. Hazod, Stetige Faltungshalbgruppen von Wahrscheinlichkeitsmaßen und erzeugende Distributionen. XIII, 157 Seiten. 1977.

Vol. 596: K. Deimling, Ordinary Differential Equations in Banach Spaces. VI, 137 pages. 1977.

Vol. 597: Geometry and Topology, Rio de Janeiro, July 1976. Proceedings. Edited by J. Palis and M. do Carmo. VI, 866 pages. 1977.

Vol. 598: J. Hoffmann-Jørgensen, T. M. Liggett et J. Neveu, Ecole d'Eté de Probabilités de Saint-Flour VI – 1976. Edité par P.-L. Hennequin. XII, 447 pages. 1977.

Vol. 599: Complex Analysis, Kentucky 1976. Proceedings. Edited by J. D. Buckholtz and T. J. Suffridge. X, 159 pages. 1977.

Vol. 600: W. Stoll, Value Distribution on Parabolic Spaces. VIII, 216 pages. 1977.

Vol. 601: Modular Functions of one Variable V, Bonn 1976. Proceedings. Edited by J.-P. Serre and D. B. Zagier. VI, 294 pages. 1977.

Vol. 602: J. P. Brezin, Harmonic Analysis on Compact Solvmanifolds. VIII, 179 pages. 1977.

Vol. 603: B. Moishezon, Complex Surfaces and Connected Sums of Complex Projective Planes. IV, 234 pages. 1977.

Vol. 604: Banach Spaces of Analytic Functions, Kent, Ohio 1976. Proceedings. Edited by J. Baker, C. Cleaver and Joseph Diestel. VI, 141 pages. 1977.

Vol. 605: Sario et al., Classification Theory of Riemannian Manifolds. XX, 498 pages. 1977.

Vol. 606: Mathematical Aspects of Finite Element Methods. Proceedings 1975. Edited by I. Galligani and E. Magenes. VI, 362 pages. 1977.

Vol. 607: M. Métivier, Reelle und Vektorwertige Quasimartingale und die Theorie der Stochastischen Integration. X, 310 Seiten. 1977.

Vol. 608: Bigard et al., Groupes et Anneaux Réticulés. XIV, 334 pages. 1977.

Vol. 609: General Topology and Its Relations to Modern Analysis and Algebra IV. Proceedings 1976. Edited by J. Novák. XVIII, 225 pages. 1977.

Vol. 610: G. Jensen, Higher Order Contact of Submanifolds of Homogeneous Spaces. XII, 154 pages. 1977.

Vol. 611: M. Makkai and G. E. Reyes, First Order Categorical Logic. VIII, 301 pages. 1977.

Vol. 612: E. M. Kleinberg, Infinitary Combinatorics and the Axiom of Determinateness. VIII, 150 pages. 1977.

Vol. 613: E. Behrends et al., L^p-Structure in Real Banach Spaces. X, 108 pages. 1977.

Vol. 614: H. Yanagihara, Theory of Hopf Algebras Attached to Group Schemes. VIII, 308 pages. 1977.

Vol. 615: Turbulence Seminar, Proceedings 1976/77. Edited by P. Bernard and T. Ratiu. VI, 155 pages. 1977.

Vol. 616: Abelian Group Theory, 2nd New Mexico State University Conference, 1976. Proceedings. Edited by D. Arnold, R. Hunter and E. Walker. X, 423 pages. 1977.

Vol. 617: K. J. Devlin, The Axiom of Constructibility: A Guide for the Mathematician. VIII, 96 pages. 1977.

Vol. 618: I. I. Hirschman, Jr. and D. E. Hughes, Extreme Eigen Values of Toeplitz Operators. VI, 145 pages. 1977.

Vol. 619: Set Theory and Hierarchy Theory V, Bierutowice 1976. Edited by A. Lachlan, M. Srebrny, and A. Zarach. VIII, 358 pages. 1977.

Vol. 620: H. Popp, Moduli Theory and Classification Theory of Algebraic Varieties. VIII, 189 pages. 1977.

Vol. 621: Kauffman et al., The Deficiency Index Problem. VI, 112 pages. 1977.

Vol. 622: Combinatorial Mathematics V, Melbourne 1976. Proceedings. Edited by C. Little. VIII, 213 pages. 1977.

Vol. 623: I. Erdelyi and R. Lange, Spectral Decompositions on Banach Spaces. VIII, 122 pages. 1977.

Vol. 624: Y. Guivarc'h et al., Marches Aléatoires sur les Groupes de Lie. VIII, 292 pages. 1977.

Vol. 625: J. P. Alexander et al., Odd Order Group Actions and Witt Classification of Innerproducts. IV, 202 pages. 1977.

Vol. 626: Number Theory Day, New York 1976. Proceedings. Edited by M. B. Nathanson. VI, 241 pages. 1977.

Vol. 627: Modular Functions of One Variable VI, Bonn 1976. Proceedings. Edited by J.-P. Serre and D. B. Zagier. VI, 339 pages. 1977.

Vol. 628: H. J. Baues, Obstruction Theory on the Homotopy Classification of Maps. XII, 387 pages. 1977.

Vol. 629: W.A. Coppel, Dichotomies in Stability Theory. VI, 98 pages. 1978.

Vol. 630: Numerical Analysis, Proceedings, Biennial Conference, Dundee 1977. Edited by G. A. Watson. XII, 199 pages. 1978.

Vol. 631: Numerical Treatment of Differential Equations. Proceedings 1976. Edited by R. Bulirsch, R. D. Grigorieff, and J. Schröder. X, 219 pages. 1978.

Vol. 632: J.-F. Boutot, Schéma de Picard Local. X, 165 pages. 1978.

Vol. 633: N. R. Coleff and M. E. Herrera, Les Courants Résiduels Associés à une Forme Méromorphe. X, 211 pages. 1978.

Vol. 634: H. Kurke et al., Die Approximationseigenschaft lokaler Ringe. IV, 204 Seiten. 1978.

Vol. 635: T. Y. Lam, Serre's Conjecture. XVI, 227 pages. 1978.

Vol. 636: Journées de Statistique des Processus Stochastiques, Grenoble 1977, Proceedings. Edité par Didier Dacunha-Castelle et Bernard Van Cutsem. VII, 202 pages. 1978.

Vol. 637: W. B. Jurkat, Meromorphe Differentialgleichungen. VII, 194 Seiten. 1978.

Vol. 638: P. Shanahan, The Atiyah-Singer Index Theorem, An Introduction. V, 224 pages. 1978.

Vol. 639: N. Adasch et al., Topological Vector Spaces. V, 125 pages. 1978.

Vol. 640: J. L. Dupont, Curvature and Characteristic Classes. X, 175 pages. 1978.

Vol. 641: Séminaire d'Algèbre Paul Dubreil, Proceedings Paris 1976–1977. Edité par M. P. Malliavin. IV, 367 pages. 1978.

Vol. 642: Theory and Applications of Graphs, Proceedings, Michigan 1976. Edited by Y. Alavi and D. R. Lick. XIV, 635 pages. 1978.

Vol. 643: M. Davis, Multiaxial Actions on Manifolds. VI, 141 pages. 1978.

Vol. 644: Vector Space Measures and Applications I, Proceedings 1977. Edited by R. M. Aron and S. Dineen. VIII, 451 pages. 1978.

Vol. 645: Vector Space Measures and Applications II, Proceedings 1977. Edited by R. M. Aron and S. Dineen. VIII, 218 pages. 1978.

Vol. 646: O. Tammi, Extremum Problems for Bounded Univalent Functions. VIII, 313 pages. 1978.

Vol. 647: L. J. Ratliff, Jr., Chain Conjectures in Ring Theory. VIII, 133 pages. 1978.

Vol. 648: Nonlinear Partial Differential Equations and Applications, Proceedings, Indiana 1976–1977. Edited by J. M. Chadam. VI, 206 pages. 1978.

Vol. 649: Séminaire de Probabilités XII, Proceedings, Strasbourg, 1976–1977. Edité par C. Dellacherie, P. A. Meyer et M. Weil. VIII, 805 pages. 1978.

Vol. 650: C*-Algebras and Applications to Physics. Proceedings 1977. Edited by H. Araki and R. V. Kadison. V, 192 pages. 1978.

Vol. 651: P. W. Michor, Functors and Categories of Banach Spaces. VI, 99 pages. 1978.

Vol. 652: Differential Topology, Foliations and Gelfand-Fuks-Cohomology, Proceedings 1976. Edited by P. A. Schweitzer. XIV, 252 pages. 1978.

Vol. 653: Locally Interacting Systems and Their Application in Biology. Proceedings, 1976. Edited by R. L. Dobrushin, V. I. Kryukov and A. L. Toom. XI, 202 pages. 1978.

Vol. 654: J. P. Buhler, Icosahedral Golois Representations. III, 143 pages. 1978.

Vol. 655: R. Baeza, Quadratic Forms Over Semilocal Rings. VI, 199 pages. 1978.

Vol. 656: Probability Theory on Vector Spaces. Proceedings, 1977. Edited by A. Weron. VIII, 274 pages. 1978.

Vol. 657: Geometric Applications of Homotopy Theory I, Proceedings 1977. Edited by M. G. Barratt and M. E. Mahowald. VIII, 459 pages. 1978.

Vol. 658: Geometric Applications of Homotopy Theory II, Proceedings 1977. Edited by M. G. Barratt and M. E. Mahowald. VIII, 487 pages. 1978.

Vol. 659: Bruckner, Differentiation of Real Functions. X, 247 pages. 1978.

Vol. 660: Equations aux Dérivée Partielles. Proceedings, 1977. Edité par Pham The Lai. VI, 216 pages. 1978.

Vol. 661: P. T. Johnstone, R. Paré, R. D. Rosebrugh, D. Schumacher, R. J. Wood, and G. C. Wraith, Indexed Categories and Their Applications. VII, 260 pages. 1978.

Vol. 662: Akin, The Metric Theory of Banach Manifolds. XIX, 306 pages. 1978.

Vol. 663: J. F. Berglund, H. D. Junghenn, P. Milnes, Compact Right Topological Semigroups and Generalizations of Almost Periodicity. X, 243 pages. 1978.

Vol. 664: Algebraic and Geometric Topology, Proceedings, 1977. Edited by K. C. Millett. XI, 240 pages. 1978.

Vol. 665: Journées d'Analyse Non Linéaire. Proceedings, 1977. Edité par P. Bénilan et J. Robert. VIII, 256 pages. 1978.

Vol. 666: B. Beauzamy, Espaces d'Interpolation Réels: Topologie et Géometrie. X, 104 pages. 1978.

Vol. 667: J. Gilewicz, Approximants de Padé. XIV, 511 pages. 1978.

Vol. 668: The Structure of Attractors in Dynamical Systems. Proceedings, 1977. Edited by J. C. Martin, N. G. Markley and W. Perrizo. VI, 264 pages. 1978.

Vol. 669: Higher Set Theory. Proceedings, 1977. Edited by G. H. Müller and D. S. Scott. XII, 476 pages. 1978.

Vol. 670: Fonctions de Plusieurs Variables Complexes III, Proceedings, 1977. Edité par F. Norguet. XII, 394 pages. 1978.

Vol. 671: R. T. Smythe and J. C. Wierman, First-Passage Perculation on the Square Lattice. VIII, 196 pages. 1978.

Vol. 672: R. L. Taylor, Stochastic Convergence of Weighted Sums of Random Elements in Linear Spaces. VII, 216 pages. 1978.

Vol. 673: Algebraic Topology, Proceedings 1977. Edited by P. Hoffman, R. Piccinini and D. Sjerve. VI, 278 pages. 1978.

Vol. 674: Z. Fiedorowicz and S. Priddy, Homology of Classical Groups Over Finite Fields and Their Associated Infinite Loop Spaces. VI, 434 pages. 1978.

Vol. 675: J. Galambos and S. Kotz, Characterizations of Probability Distributions. VIII, 169 pages. 1978.

Vol. 676: Differential Geometrical Methods in Mathematical Physics II, Proceedings, 1977. Edited by K. Bleuler, H. R. Petry and A. Reetz. VI, 626 pages. 1978.

Vol. 677: Séminaire Bourbaki, vol. 1976/77, Exposés 489–506. IV, 264 pages. 1978.

Vol. 678: D. Dacunha-Castelle, H. Heyer et B. Roynette. Ecole d'Eté de Probabilités de Saint-Flour. VII-1977. Edité par P. L. Hennequin. IX, 379 pages. 1978.

Vol. 679: Numerical Treatment of Differential Equations in Applications, Proceedings, 1977. Edited by R. Ansorge and W. Törnig. IX, 163 pages. 1978.

Vol. 680: Mathematical Control Theory, Proceedings, 1977. Edited by W. A. Coppel. IX, 257 pages. 1978.

Vol. 681: Séminaire de Théorie du Potentiel Paris, No. 3, Directeurs: M. Brelot, G. Choquet et J. Deny. Rédacteurs: F. Hirsch et G. Mokobodzki. VII, 294 pages. 1978.

Vol. 682: G. D. James, The Representation Theory of the Symmetric Groups. V, 156 pages. 1978.

Vol. 683: Variétés Analytiques Compactes, Proceedings, 1977. Edité par Y. Hervier et A. Hirschowitz. V, 248 pages. 1978.

Vol. 684: E. E. Rosinger, Distributions and Nonlinear Partial Differential Equations. XI, 146 pages. 1978.

Vol. 685: Knot Theory, Proceedings, 1977. Edited by J. C. Hausmann. VII, 311 pages. 1978.

Vol. 686: Combinatorial Mathematics, Proceedings, 1977. Edited by D. A. Holton and J. Seberry. IX, 353 pages. 1978.

Vol. 687: Algebraic Geometry, Proceedings, 1977. Edited by L. D. Olson. V, 244 pages. 1978.

Vol. 688: J. Dydak and J. Segal, Shape Theory. VI, 150 pages. 1978.

Vol. 689: Cabal Seminar 76–77, Proceedings, 1976–77. Edited by A.S. Kechris and Y. N. Moschovakis. V, 282 pages. 1978.

Vol. 690: W. J. J. Rey, Robust Statistical Methods. VI, 128 pages. 1978.

Vol. 691: G. Viennot, Algèbres de Lie Libres et Monoïdes Libres. III, 124 pages. 1978.

Vol. 692: T. Husain and S. M. Khaleelulla, Barrelledness in Topological and Ordered Vector Spaces. IX, 258 pages. 1978.

Vol. 693: Hilbert Space Operators, Proceedings, 1977. Edited by J. M. Bachar Jr. and D. W. Hadwin. VIII, 184 pages. 1978.

Vol. 694: Séminaire Pierre Lelong – Henri Skoda (Analyse) Année 1976/77. VII, 334 pages. 1978.

Vol. 695: Measure Theory Applications to Stochastic Analysis, Proceedings, 1977. Edited by G. Kallianpur and D. Kölzow. XII, 261 pages. 1978.

Vol. 696: P. J. Feinsilver, Special Functions, Probability Semigroups, and Hamiltonian Flows. VI, 112 pages. 1978.

Vol. 697: Topics in Algebra, Proceedings, 1978. Edited by M. F. Newman. XI, 229 pages. 1978.

Vol. 698: E. Grosswald, Bessel Polynomials. XIV, 182 pages. 1978.

Vol. 699: R. E. Greene and H.-H. Wu, Function Theory on Manifolds Which Possess a Pole. III, 215 pages. 1979.